农林类高职高专基础课系列教材

English for Agriculture Careers

农业职业英语

主　编　杜文贤　王立阁
副主编　姚　阳　胡月红　苏　波　刘　波
参　编　李爱华　冷　学　陈诗宇　王丽君
刘立英　王永淇　刘来毅　张海涛
杨振野　王　爽

English for Agriculture Careers

English for Agriculture Careers

English for Agriculture Careers

北京理工大学出版社
BEIJING INSTITUTE OF TECHNOLOGY PRESS

内 容 提 要

《农业职业英语》是一本涵盖校园生活、电子商务、食品安全与加工、生态旅游、酒店服务、现代农机、现代种植业、现代养殖业、环境规划和求职等诸多主题的、公共英语和专业英语巧妙融合的教材。

本着“基于科普、适度深耕、适当模糊专业界限、体现农业特色”的编写宗旨，在对未来就业岗位对农业类专业毕业生的英语要求和学生的英语基础、学习习惯和接受能力进行了全面深入的调研和听取专业资深教师的意见的基础上，本着“实用、够用、应用”的原则，编者打造了这部具有领先示范效应的、内容丰富全面的英语教材。

图书在版编目（CIP）数据

农业职业英语 / 杜文贤，王立阁主编 .—北京：北京理工大学出版社，2021.1（2022.8 重印）

ISBN 978-7-5682-8996-2

Ⅰ . ①农…　Ⅱ . ①杜…　②王…　Ⅲ . ①农业—英语—高等学校—教材　Ⅳ . ① S

中国版本图书馆 CIP 数据核字（2020）第 164922 号

出版发行 / 北京理工大学出版社有限责任公司
社　　址 / 北京市海淀区中关村南大街5号
邮　　编 / 100081
电　　话 /（010）68914775（总编室）
（010）82562903（教材售后服务热线）
（010）68944723（其他图书服务热线）
网　　址 / http：//www.bitpress.com.cn
经　　销 / 全国各地新华书店
印　　刷 / 河北鑫彩博图印刷有限公司
开　　本 / 889毫米×1194毫米　1/16
印　　张 / 10
字　　数 / 223千字
版　　次 / 2021年1月第1版　　2022年8月第4次印刷
定　　价 / 49.80元

责任编辑 / 阎少华
文案编辑 / 阎少华
责任校对 / 周瑞红
责任印制 / 边心超

前言
FOREWORD

本书是一本包含若干主题的、公共英语和专业英语巧妙融合的英语教材，可满足农业类院校几乎所有专业的英语教学，内容丰富全面，具有极强的实用性。

适用对象

本书适合作为农业类院校学生教材，也可作为相关从业人员自学工具书使用。

教材结构

本套教材包括主修教材和辅修教材两册。主修教材满足课堂使用，辅修教材供课后以及实习实训期间巩固复习之用。教材电子课件及习题答案将适时上传至出版社网站，供使用者学习和参考。

教材内容

本书设置十一个单元，每单元一个主题，即校园生活、电子商务、食品安全、食品加工、生态旅游、酒店服务、现代农机、现代种植业、现代养殖业、环境规划和求职。

每单元由五个模块组成：

Task 1　First Sight　预热

通过图片与关键词匹配，引入主题，并通过词汇搭配展开话题讨论，为主题内容的学习做好铺垫和引领。

Task 2　Better Acquaintance　口语交际

一个示范对话领先，一个操练对话紧随其后。五个替换练习旨在扩充本单元主题下相关知识点的词汇量储备。迷你小对话匹配题注重从不同角度、不同知识点切入话题并展开交流，实现了交际场景的转换和拓宽。

Task 3　Further Development　阅读及语言知识

两篇文章内容紧扣单元主题。Passage 1所附习题侧重词汇、语言点、句子结构及翻译训练，Passage 2则侧重阅读能力训练。

Task 4　Related Information　知识延伸

作为单元主题内容的有力补充，本版块语言精练，所附插图直观生动，内容实用且操作性强。既可作为补充信息在课堂上学习讨论，也可作为自学内容课后完成。

Task 5　Pop Quiz　随堂测验

通过词汇匹配、选择和段落填空三个小版块实现对课堂所学知识的即学即测，从而对教学效果做出评估。

教材特色

在众多院校中，农业类院校以其学生基础总体薄弱、涉农专业、课程设置和注重动手实践能力而显得与众不同。事实证明，单纯的公共英语（EGP）教学已不能满足未来就业岗位的需求，专业英语（ESP）教学的开展势在必行。但农业类院校学生英语的短板以及专业英语中的词汇和长难句都是障碍，因此，在前期充分调研的基础上，国际教育学院多位英语教师为农业类院校学生量身定做了这部公共英语补齐短板、专业英语涉猎适度、科普与深耕相得益彰、实用性极强的教材。

本书具有以下创新：

1. 主题覆盖面广

本书设置十一个单元，全面覆盖了农业类院校相关专业。未来的实际教学中，教师可根据专业而有侧重地对教材加以选用。

2. 主、辅修教材主题一致

主修供课堂使用，辅修教材作为习题集主要供课下和学习实训期间使用。主、辅修教材主题一致，辅修教材为主修教材的补充。

3. 素材规范

采用的素材均来自国内外正规英文网站，权威性强，文字使用规范且紧跟时代。

4. 对话原创、真实

示范对话、练习对话以及迷你对话全部为编写者自主创作，根据实际需要设置场景、人物关系和对话内容，过渡自然，衔接有序，长度、难度和知识点数量都控制得恰到好处。

5. 公共英语和专业英语完美融合

单元主题的拟定充分体现了教材的编写宗旨，即“基于科普、适度深耕、适当模糊专业界限、体现农业特色”。在保证专业英语适度、够用的基础上，又兼顾弥补公共英语短板，实现公共英语和专业英语的巧妙融合。

6. 基于不同层次的难度控制

每单元习题的设置呈阶梯渐进式，基础题、中等难度题和拔高题可满足不同基础学生的需求，学生的各项英语技能都能在原有基础上得到不同程度的提高。

7. A、B级题型设置

主、辅修教材习题按照英语A、B级题型设置，使学生在日常学习中熟悉和适应考试题型。

8. 图文并茂、直观生动

选取的图片生动、恰当，趣味性强，有利于激发学生的学习兴趣。

9. 全员参与，严把教材质量关

每单元除了主编审阅外，还经过至少两位编者审阅，编者互审机制保证了教材质量。

本书由辽宁农业职业技术学院国际教育学院部分教师以及相关系部专业教师共同编写。编写分工如下：

杜文贤编写Unit 7 & Unit 8，负责与各专业的对接、沟通和调研工作；参与所有单元主题的确立；设置单元内版块、撰写模板、编者分工以及全书的统筹、审阅。

王立阁负责外联、沟通、协调；对教材的编写给予方向性指导；主导所有单元主题的确立；对素材长度、难度制订标准；参与单元版块的设计、论证、筛选及确立；人员安排。姚阳编写Unit 4 & Unit 5，胡月红编写Unit 2 & Unit 3，苏波编写Unit 1 & Unit 9，刘波编写Unit 10，李爱华编写Unit 11，冷学编写Unit 6。陈诗宇、杨振野、王爽协助对书稿进行校对、排版以及插图的搜集、编辑。王丽君、刘立英、王永淇、刘来毅和张海涛五位老师对编写范围给予指导，并对素材的难度、长度加以把关。

由于编者水平有限，时间仓促，不当和疏漏之处在所难免，恳请使用者给予批评指正，使本书能为农业类院校英语教学做出贡献。

编　者

目 录
CONTENTS

UNIT 1 Campus Life

Task 1 First Sight

1. Match the following pictures with the key words given.

A. lecture period
B. sporting activities
C. dormitory life
D. graduation speech

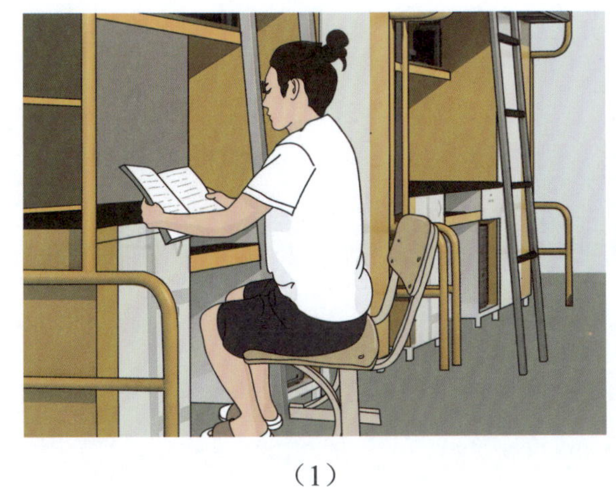

(1)

(2)

(3)

(4)

2. Answer the question by matching the following items.

What are the reasons for taking part in student clubs and organizations?

□1）make	A. your life
□2）improve	B. physical exercises
□3）do	C. extracurricular knowledge
□4）learn	D. new friends
□5）enjoy	E. communication ability

Task 2 Better Acquaintance

Conversation

Talking about Campus Life

（Jack and Mary are talking about their campus life.）

Jack: Are you enjoying your time at university?

Mary: It's okay, but not as much as I expected.

Jack: Really? Why not?

Mary:I thought that the subjects would be easy, and I would have lots of time to play, but actually I have many assignments.

Jack: What kind of assignments are you having problems with?

Mary:I find my accounting assignments very difficult.

Jack: I suggest that you turn to your professors and classmates for help.

Mary: Sounds great.

Jack: Have you joined a sports team?

Mary: Yes, I'm playing for the university volleyball team.

Jack: So am I!

Mary:I've joined the university volleyball club, and they've put me in the women's league.

Jack: I guess that's why I haven't seen you for a long time. I'm playing in a mixed league.

Mary: Cool! That sounds like a nice way to meet some new friends. What else are you doing at university apart from studying and playing volleyball?

Jack: I've also joined the university debating team, and we have a competition every month.

Mary: Wonderful. Can I join the debating team?

Jack: Of course. New people are welcome at any time.

Mary: Thanks. That's good to know. I think I'll go and sign up next week.

New Words

campus	[ˈkæmpəs]	*n.*	（大学、学院的）校园
actually	[ˈæktʃuəli]	*adv.*	实际上
assignment	[əˈsaɪnmənt]	*n.*	任务；作业
accounting	[əˈkaʊntɪŋ]	*n.*	会计
professor	[prəˈfesə(r)]	*n.*	教授
league	[liːg]	*n.*	联赛；社团
debate	[dɪˈbeɪt]	*v.*	辩论
competition	[ˌkɒmpəˈtɪʃn]	*n.*	比赛；竞赛

Phrases & Expressions

have problems with...	在……方面有问题
turn to...for help	向……寻求帮助
apart from	除了……外（还）
sign up	报名（参加课程）

Projects

Ⅰ. Substitution drills

1. Tips for better grades include

发现你的学习风格.
优化你的学习空间.
掌握重要的学习技巧.
改掉不良的学习习惯.
理解不同的测试类型.

2. College students who take part-time jobs can

丰富个人的工作经历.
了解周围的社会环境.
赚得额外的收入.
拓宽专业知识面.
合理支配时间.

3. College students should make more friends by

与室友和谐相处.
主动和新同学交流.
参加俱乐部活动.
加入学习小组.
提升自身的素质.

4. College students should learn financial management by

做好预算并记账.
精打细算旅行.
享受学生折扣.
获得奖学金.
理性购物.

5. College students' career planning should focus on

个人兴趣.
学习能力.
专业发展.
社会需求.
家庭因素.

Ⅱ. The following are questions and answers about campus life between two students. Join the questions 1-8 to the answers A-H. Then act out some of the mini-dialogs with your partner.

1. Did you have any work experience during college? ________
2. Excuse me, I am a newcomer here. Could you tell me where the school library is? ________
3. How often do you have English lessons? ________
4. How long can I keep the books? ________
5. It's a lovely day today, isn't it? ________
6. I heard you've got a new roommate. What's he like? ________
7. So you're going to graduate from your university next year. Do you have any plans for the future? ________
8. You look a bit pale. What's the matter with you, Mary? ________

A. Yes, it is. It's so pleasant after the bad weather we've been having. How about playing basketball?

B. Yeah. I am considering looking for a job as a salesman in a big company. What about you?

C. Once a week, but I try to practice it every day.

D. For two weeks. After that you must renew the books if you wish to keep them longer.

E. No problem. Go straight ahead till you come to the corner, turn left there and it's at the first turning on the right.

F. I'm afraid I've got a terrible cold.

G. Yeah. Bob moved in last week. He is a nice guy.

H. Of course I did. I did a part-time job as an English tutor.

Ⅲ. Fill in the blanks in the conversation by translating the Chinese into appropriate English.

A Good Method to Learn Chinese

（Jack and his friend Katherine are talking about a good method to learn Chinese.）

Katherine: Are you OK? You look pale.

Jack: I'm fine, Katherine. I （1） ________________ （昨晚熬夜） working on my Chinese.

Katherine: Oh, don't worry. You know what? （2） ________________ （我发现一个学汉语的新方法） and it works very well.

Jack: You don't say! Do tell me. I've spent all these months trying to learn something new about it, but I've made little progress!

Katherine: Sure. （3） ________________ （我的方法就是“唱歌学中文”）.

Jack: So you're learning Chinese songs?

Katherine: You can say that. I can actually sing some of them now.

Jack: That's a real achievement. How did you do that?

Katherine: I started by listening to a song again and again and after trying several times I was able to follow the singer.

Jack: （4） ________________ （挺有道理的）.

Katherine: So it does. I've learned some Chinese folk songs this way. They are clearly presented and easy to follow.

Jack: Can I join you, Katherine? （5） ________________ （我也想试试这个方法）.

Katherine: Why not? I downloaded a beautiful song the other day. We can learn it together now.

Jack: Great. Do you think we need a Chinese dictionary?

Katherine: Yes, just in case we run into new characters.

Jack: OK. I will bring it with me.

Ⅳ. Remember the useful expressions concerning campus life.

1. Which department are you in and what is your major?
2. How is your English lesson going?
3. Excuse me, could you tell me the way to the chemistry laboratory?
4. What shall I do when I check out these books?
5. Do you get along well with your roommates?
6. Do you have a part-time job to support yourself through college?
7. What is the topic of your paper?
8. What do you think of studying abroad?

Task 3 Further Development

Passage 1

Pre-reading Task

Read the following statements and tick in the columns under "True", "False" or "Unsure".

Statements	True	False	Unsure
Having goals in college is very essential.			
College life is always difficult and stressful.			
You can set both large and small goals in your college life.			
The more specific your goal is, the more realistic it is.			
It is easy for the most driven and determined college students to set goals.			

How to Set College Goals

Having goals in college can be a great way to stay focused, motivate yourself, and keep your priorities in order when things get stressful and overwhelming. But just how can you set your college goals in a way that sets you up for success?

Think about Your End Goals

What kind of goals do you want to achieve during your time in school? These goals can be large or small. Having the main goal in mind is the first, and perhaps most important step, in setting realistic goals.

Be Specific with Your Goals

Instead of "Do better in Chemistry," set your goal as "Earn at least a B in Chemistry this term." Being as specific as possible while setting your goals can help make your goals as realistic as possible—meaning you'll be more likely to achieve them.

Be Realistic about Your Goals

If you barely passed most of your classes last semester and are now on academic probation, setting a goal of earning a 4.0 next semester is probably unrealistic. Spend some time thinking about what makes sense for you as a learner, as a student, and as a person.

Think about Your Personal and Intellectual Strengths

Setting goals can be challenging for even the most driven and determined college students. If you set yourself up to do things that are a bit too challenging, however, you can end up setting yourself up for failure instead of for success. Spend some time thinking about your own personal and intellectual strengths. In essence, use your strengths to find ways to overcome your weaknesses.

New Words

focus	[ˈfəʊkəs]	*v.*	集中（注意力、精力）
motivate	[ˈməʊtɪveɪt]	*v.*	激励
priority	[praɪˈɒrəti]	*n.*	首要事情
overwhelming	[ˌəʊvəˈwelmɪŋ]	*adj.*	无法抗拒的
achieve	[əˈtʃiːv]	*v.*	完成
realistic	[ˌriːəˈlɪstɪk]	*adj.*	现实的
specific	[spəˈsɪfɪk]	*adj.*	具体的
barely	[ˈbeəli]	*adv.*	仅仅；刚刚
semester	[sɪˈmestə(r)]	*n.*	学期
academic	[ˌækəˈdemɪk]	*adj.*	学术的
probation	[prəˈbeɪʃn]	*n.*	试读期
sense	[sens]	*n.*	道理
personal	[ˈpɜːsənl]	*adj.*	个人的
intellectual	[ˌɪntəˈlektʃuəl]	*adj.*	有才智的
strength	[streŋθ]	*n.*	长处
challenging	[ˈtʃælɪndʒɪŋ]	*adj.*	挑战性的
driven	[ˈdrɪvn]	*adj.*	奋发努力的
determined	[dɪˈtɜːmɪnd]	*adj.*	坚定的
essence	[ˈesns]	*n.*	本质
overcome	[ˌəʊvəˈkʌm]	*v.*	克服
weakness	[ˈwiːknəs]	*n.*	弱点

Phrases & Expressions

set...up	为……做准备
make sense	有道理
end up	以……告终
in essence	实质上

Projects

Ⅰ. Fill in the blanks with the proper words given, changing the form if necessary.

priority	achieve	challenge	focus
realistic	stress	strength	personal

1. We have __________ what we set out to do.
2. I have had a __________ and rewarding career as a teacher.
3. You can stay calm, __________ and ready to answer questions in winning ways.
4. Many people complain that their jobs are uninteresting and __________.
5. It may take a few weeks for you to build up your __________ again.
6. The important thing is to have __________ expectations about what you can and can't accomplish.
7. The manager said that he had resigned for __________ reasons.
8. The search for a new vaccine will take __________ over all other medical research.

Ⅱ. Translate the following sentences by imitating the examples given, paying attention to the underlined phrases or structures.

1. Having the main goal in mind is the first, and perhaps most important step, in setting realistic goals.
   收集信息对于商人来说非常重要。
   ____.
2. Being as specific as possible while setting your goals can help make your goals as realistic as possible—meaning you'll be more likely to achieve them.
   尽管他们失败了，但他们仍像以前一样努力工作。
   ____.
3. Spend some time thinking about what makes sense for you as a learner, as a student, and as a person.
   他们花了两年时间造这座桥。
   ____.
4. If you set yourself up to do things that are a bit too challenging, however, you can end up setting yourself up for failure instead of for success.

如果你不努力学习，你最终就会考试不及格。

____.

5. In essence, use your strengths to find ways to overcome your weaknesses.
所有的计算机其实都一样。

____.

Passage 2

Self-growth in College

You are in college now and college is one of the most exciting stages of one's life. Close friends contribute to your self-growth, for they provide you with moral support that is so important to survive the stressful college life. Friends can lend a helping hand when necessary like collecting your homework when you're too sick to leave your dorm, and help you develop the right attitude by pointing out your weaknesses.

Being home most of your life and then suddenly finding yourself on your own in a large campus without your parents can be annoying. However, rather than thinking about your missing home too much, why not regard this new stage in your life as an opportunity for self-growth and develop the right attitude that will prepare you for the post-graduation life?

Clubs or organizations are great for self-growth, too. Here, not only do you get the opportunity to meet with likeminded people, you can also discover more things about your field of interest. You should try to build a good name in the organization by being respectful and considerate to others.

Actively participating in a class debate or lecture contributes to your self-growth as it helps you build confidence in speaking up and improves your communication skills.

New Words

moral	[ˈmɒrəl]	*adj.*	道德上的
survive	[səˈvaɪv]	*v.*	挺过；存活
annoying	[əˈnɔɪɪŋ]	*adj.*	使烦恼的
considerate	[kənˈsɪdərət]	*adj.*	考虑周到的
confidence	[ˈkɒnfɪdəns]	*n.*	自信
communication	[kəˌmjuːnɪˈkeɪʃn]	*n.*	交际

Phrases & Expressions

contribute to	有助于……
participate in	参加

Check your understanding

Choose the correct answers according to the information given in the passage.

1. According to the first paragraph, how can college students develop a positive attitude?
   A. By focusing on their studies.
   B. By taking part-time jobs.
   C. By making friends with others.
2. According to the text, college students should do many things EXCEPT________.
   A. forget the family
   B. be respectful at club meetings
   C. take an active part in a class debate
3. We can learn from the text that________.
   A. college life is very complicated
   B. taking part in a class debate or lecture contributes to your self-growth
   C. students should choose words carefully on any occasion
4. What does "likeminded"mean in the third paragraph?
   A. Kind and friendly.
   B. Having similar ideas, opinions and interests.
   C. Considerate.
5. The purpose of the text is to________.
   A. introduce how to show yourself in college
   B. tell college students how to socialize with others
   C. give college students some advice on self-growth

Task 4 Related Information

Understand the information about college meal plans.

College Meal Plans

Essentially, a meal plan is a pre-paid account for your on-campus meals. At the start of the term, you pay for all the meals you'll eat in the dining halls. You'll then use your student ID or a special meal card every time you enter a dining area, and the value of your meal will be deducted from your account.

Most colleges require residential students to get a meal plan. This is especially true for first-year students.

The price of meal plans will vary significantly from school to school. Options ranging from 7 to 21 meals a week may be available.

At most schools, your meal card will work at all dining facilities on campus giving you a wide range of options.

At some schools, the money for unused meals can be spent at a campus convenience store or even with local merchants.

Task 5 Pop Quiz

Ⅰ. Working with words

Match the following words and phrases with their Chinese equivalents.

A. dormitory life
B. graduation paper
C. academic practice
D. psychological adjustment
E. interpersonal relationship
F. tuition
G. scholarship
H. entrepreneurship of college students
I. extracurricular activities
J. compulsory courses
K. career planning
L. learning resources

1. 学习资源（ ）	6. 毕业论文（ ）
2. 人际关系（ ）	7. 心理调适（ ）
3. 课外活动（ ）	8. 寝室生活（ ）
4. 职业规划（ ）	9. 大学生创业（ ）
5. 必修课（ ）	10. 学术实践（ ）

Ⅱ. Multiple choice

1. The term "________" is used loosely to denote not just dorm status, but distance.

A. college student　B. commuter student　C. sophomore

2. Examine and evaluate your ________ and decide how you can improve your study habits by tapping into your personal strengths.

A. eating habit　B. learning place　C. learning style

3. The ________ is a way of life for most college students, with research showing them to be one of the most connected groups.

A. vehicle　B. dinner　C. internet

4. If you're wondering how to ________ school and personal life while you're in college, it all starts with creating a schedule for yourself.

A. choose B. compare C. balance

5. Living on campus definitely has its ________. You get to live among your fellow students and making it to class on time is as simple as walking across campus.

A. disadvantages B. benefits C. weaknesses

6. The first few weeks away from home are no doubt the hardest for many freshmen, for they may be ________ sometimes.

A. interested B. excited C. homesick

7. In order to graduate, a student must attend a certain number of ________, which gives him a ________ he may count towards a degree.

A. course; credit B. courses; credit C. activities; grade

8. Don't wait until you've graduated to begin preparing for your job search. Becoming an attractive candidate begins from the moment ________.

A. you start college B. you leave college C. you finish study

Ⅲ. Fill in the blanks with the proper choices given.

Your college years are arguably some of the most stressful of your life. __（1）__ is to express yourself creatively. Here is how to get started:

Make time in your schedule. The first step is finding room in your schedule to actually sit down and let your creative juices flow. Everyone is busy in college. However, one of the best things you can do for yourself is __（2）__. If you want to actually make your art a part of your career though, you will have to devote a serious chunk of your day to it.

Take a course. Schedule an appointment with your adviser and see if the college you attend __（3）__ that you're interested in.

Join a club or make one. Chances are that there are other students at your school __（4）__. If there isn't one, check out the procedures for going about making one!

Constantly search for inspiration. This could mean something as simple as looking for blogs that showcase work that you admire or going to a museum in your college town and wandering around for a bit.

Even if you don't feel that you have enough talent to make your art a part of your professional career, it never hurts to try. The skills that you learn while working at your art might even possibly help you __（5）__.

A. to take time to relax and let your mind wander as it pleases

B. offers a course in the creative field

C. that share the same interests as you

D. in the workplace down the road

E. The best way to get it all out

UNIT 2 E-commerce World

Task 1 First Sight

1. Match the following pictures with the key words given.

A. cold chain B. live streaming C. warehouse D. online shopping

(1)

(2)

(3)

(4)

2. Answer the question by matching the following items.

What are the benefits of e-commerce to businesses?

□1) reduce	A. the business processes
□2) expand	B. the brand image
□3) improve	C. the cost
□4) simplify	D. better customer services
□5) provide	E. their market

Task 2 Better Acquaintance

Conversation

What Is E-commerce?

(Mary is asking Nancy about e-commerce.)

Mary: This mango is so tasty. Where did you get it?

Nancy: I bought it from a Hainan farmer online. He has been selling mangoes through live streaming for a week.

Mary: These days quite a few farmers sell their produce online. E-commerce is growing very rapidly. Nancy, do you know the definition of e-commerce?

Nancy: Yes, I do. E-commerce stands for electronic commerce. It is a type of business model that does commercial transactions through the Internet. Put simply, e-commerce is any business activity that happens online.

Mary: Can you give me some examples of e-commerce?

Nancy: Sure. Examples include online shopping, Internet banking, electronic payments, purchasing tickets online and online auctions.

Mary: Well, is the term B2C related to e-commerce?

Nancy: Yes, it is the most common type of e-commerce. B2C stands for business to consumer, which refers to transactions between online retailers and their customers.

Mary: I see. Could I have another mango?

Nancy: Of course. Help yourself.

New Words

live streaming [laɪv] [striːmɪŋ] *n.* 直播

produce	[ˈprɔdjuːs]	*n.*	农产品；产品
rapidly	[ˈræpɪdli]	*adv.*	快速地
definition	[ˌdefɪˈnɪʃn]	*n.*	定义
electronic	[ɪˌlekˈtrɒnɪk]	*adj.*	电子的
commerce	[ˈkɒmɜːs]	*n.*	贸易；商务
model	[ˈmɒdl]	*n.*	模式
commercial	[kəˈmɜːʃl]	*adj.*	商业的
transaction	[trænˈzækʃn]	*n.*	（一笔）交易
purchase	[ˈpɜːtʃəs]	*v.*	买
auction	[ˈɔːkʃn]	*n.*	拍卖
retailer	[ˈriːteɪlə]	*n.*	零售店

Phrases & Expressions

stand for	是……的缩写；代表
put simply	简单地说
be related to	与……有关
refer to	指的是

Projects

Ⅰ. Substitution drills

1. Online shopping

是电商的一部分.
是一种新的购物方式.
节省时间和金钱.
特别有吸引力.
很方便.

2. Internet banking may include

在网上缴付账单.
在网上转账.
在网上存款.
在网上查余额.
在网上设置定期付款.

3. An e-commerce platform allows business owners to	回顾店铺业绩. 管理库存. 明确产品定价. 进行促销活动. 跟踪销售.
4. B2C businesses need to know	他们的客户是谁. 客户有什么偏好. 客户有什么痛点. 客户的需求. 在哪里找到客户.
5. Some of the most powerful B2C marketing strategies include	社交媒体营销和广告. 创意大赛. 忠诚和奖励计划. 赠品和免费附加品. 影响者营销.

Ⅱ. The following are questions and answers about logistics between two friends. Join the questions 1-8 to the answers A-H. Then act out some of the mini-dialogs with your partner.

1. Hello, this is Tom Black. I'm calling to enquire if it is possible to effect shipment in June. ________
2. Could you possibly make the delivery no later than April? ________
3. How about making shipment via Hong Kong? ________
4. Could you tell me the earliest time when you can make delivery? ________
5. How long will it take you to make shipment? ________
6. What about extra transportation charges for the early shipment? ________
7. Will the cartons be strong enough for long-distance transportation? ________
8. Do you mind if I give you a suggestion about the inner packing of the products? ________

A. There will be no extra transportation charges.

B. Not at all. Go ahead.

C. Next month.

D. Of course. We use iron straps for reinforcement.

E. In June? That is one month earlier than scheduled.

F. That sounds good. We hope to receive our goods at the earliest possible time.

G. Two weeks.

H. I'm sorry we can't. Our company has been fully committed these months.

Ⅲ. Fill in the blanks in the conversation by translating the Chinese into English.

Selling Produce Through Live Streaming

(Farmer Jack and Farmer Mike both have online stores. They are talking about live streaming.)

Mike: Hi, Jack. I have 1,000 kilograms of apples to sell. What about you?

Jack: All my apples are sold out.

Mike: (1) ________________________________ (真是不可思议). How did you make it?

Jack: I sold them through live streaming.

Mike: Good for you! Why do people like (2) ___________ (这种新的销售方式) products?

Jack: Because they can see (3) ____________ (产品真实的样子), rather than just photos. At the same time, they can (4) ____________ (直接问任何问题) about the products.

Mike: I see. What do you need to do when you do live streaming?

Jack: I need to (5) ___________ (介绍特点) of apples and demonstrate how good they are. I also need to answer the viewers' questions.

Mike: Got it. Can you help me sell my apples?

Jack: Sure, I'd love to.

Ⅳ. Remember the useful expressions concerning e-commerce.

1. This mango is so tasty. Where did you get it?
2. E-commerce is growing very rapidly.
3. E-commerce stands for electronic commerce. Put simply, e-commerce is any business activity that happens online.
4. Can you give me some examples of e-commerce?
5. Examples of e-commerce include online shopping, Internet banking, electronic payments, purchasing tickets online and online auctions.
6. Well, is the term B2C related to e-commerce?
7. B2C stands for business to consumer, which refers to transactions between online retailers and their customers.
8. I need to introduce the features of apples and demonstrate how good they are.

Task 3 Further Development

Passage 1

Pre-reading Task

Read the following statements and tick in the columns under “True”, “False” or “Unsure”.

Statements	True	False	Unsure
The e-commerce sector is already booming in urban areas thanks to the “Internet Plus Agriculture” model.			
One of the top priorities is to use communication technology in the distribution of agricultural products.			
Measures will be taken to ensure that agricultural products are safe and of high quality.			
Support facilities will be strengthened to try to improve the farm produce’s marketability and resolve the overpricing of quality products.			
Information technologies will be harnessed to try to create more efficient, digitized and smart agricultural production.			

E-commerce Sector Is Booming in Rural Areas

Huo Liang earns about 1,000 yuan（$158）a month running an online shop to sell millet, a humble but nutritious food popular among Chinese customers.

His earnings are remarkable for a financially disadvantaged family in Tongyu county, in Northeast China’s Jilin province.

Qu Dongyu, Vice Minister of Agriculture and Rural Affairs, says the e-commerce sector is already booming in rural areas thanks to the “Internet Plus Agriculture” model.

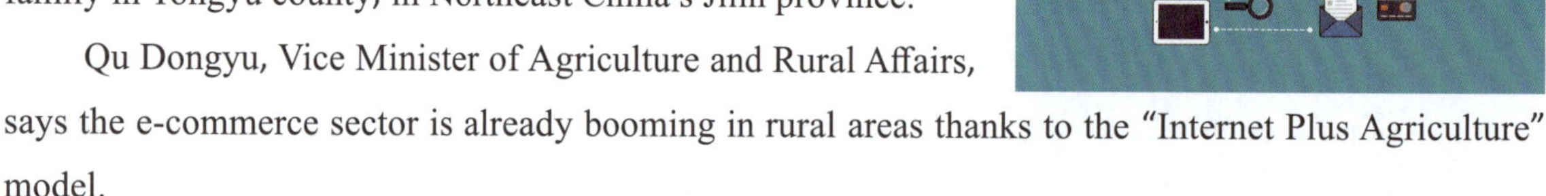

In 2017, 756 counties across the country became models for conducting e-commerce in rural areas. Online retail sales in rural areas totaled 1.25 trillion yuan, or nearly 190 billion U.S. dollars, of which sales of agricultural products approached 300 billion yuan. More than 28 million jobs were created as a result. One of the top priorities is to use information technology in the distribution of agricultural products.

A guideline will be released on widening the agricultural products’ circulation, and issues of connecting the production and marketing process will also be solved. Measures will be taken to ensure that agricultural products are safe and of high quality. Support facilities such as processing, packaging, storage, preservation and cold chain logistics will be strengthened to try to improve the farm produce’s marketability and resolve the underpricing of quality products.

Information technologies, including big data, the Internet of Things and cloud computing, will be harnessed to try to create more efficient, digitized and smart agricultural production.

Internet infrastructure in rural areas will also be improved, and agriculture-related public information and service platforms will be established as well.

Farmers will also receive training on how to better use information technology and their smart phones can become their new farm implements.

New Words

sector	[ˈsektə]	*n.*	行业
boom	[buːm]	*v.*	迅速发展
rural	[ˈrʊərəl]	*adj.*	农村的
millet	[ˈmɪlɪt]	*n.*	小米
humble	[ˈhʌmbl]	*adj.*	不起眼的
nutritious	[njuˈtrɪʃəs]	*adj.*	有营养的
remarkable	[rɪˈmɑːkəbl]	*adj.*	非凡的
financially	[faɪˈnænʃəli]	*adv.*	经济上；财政上
disadvantaged	[ˌdɪsədˈvɑːntɪdʒd]	*adj.*	弱势的
total	[ˈtəʊtl]	*v.*	总数达
trillion	[ˈtrɪljən]	*n.*	万亿
billion	[ˈbɪljən]	*n.*	十亿
approach	[əˈprəʊtʃ]	*v.*	接近
technology	[tekˈnɒlədʒi]	*n.*	技术
distribution	[ˌdɪstrɪˈbjuːʃn]	*n.*	经销；（网络）分销
guideline	[ˈgaɪdlaɪn]	*n.*	指导方针
release	[rɪˈliːs]	*v.*	发布
circulation	[ˌsɜːkjəˈleɪʃn]	*n.*	流通
processing	[ˈprəʊsesɪŋ]	*n.*	加工
packaging	[ˈpækɪdʒɪŋ]	*n.*	包装
storage	[ˈstɔːrɪdʒ]	*n.*	贮存
preservation	[ˌprezəˈveɪʃn]	*n.*	保存
strengthen	[ˈstreŋθn]	*v.*	加强
resolve	[rɪˈzɒlv]	*v.*	解决
harness	[ˈhɑːnɪs]	*v.*	利用
efficient	[ɪˈfɪʃnt]	*adj.*	效率高的
digitized	[ˈdidʒitaɪzd]	*adj.*	数字化的

infrastructure	[ˈɪnfrəstrʌktʃə(r)]	*n.*	基础设施
implement	[ˈɪmplɪmənt]	*n.*	工具

Phrases & Expressions

thanks to	由于
retail sales	零售额
as a result	结果
support facility	辅助设施
cold chain logistics	冷链物流
big data	大数据
the Internet of Things	物联网
cloud computing	云计算
as well	也

Projects

Ⅰ. Fill in the blanks with the proper words given, changing the form if necessary.

financially	resolve	total	priority
nutritious	approach	strengthen	remarkable

1. She was a truly ________ woman.
2. It is very important to choose ________ foods.
3. Oil prices have ________ their highest level for almost ten years.
4. She is still ________ dependent on her parents.
5. Her position in the party has ________ in recent years.
6. Imports ________ $1.5 billion last year.
7. Our first ________ is to improve standards.
8. Both sides met in order to try to ________ their differences.

Ⅱ. Translate the following sentences by imitating the examples given, paying attention to the underlined phrases or structures.

1. The e-commerce sector is already booming in rural areas <u>thanks to</u> the "Internet Plus Agriculture" model.

   <u>由于</u>他的帮助，我们及时完成了任务。

   ____.

2. One of the top priorities is <u>to use information technology in the distribution of agricultural products</u>.

下一步是确保全体员工参加此项活动。

____.

3. Measures will be taken to ensure that agricultural products are safe and of high quality.
为了出席会议，他比往常起得早。

____.

4. Information technologies, including big data, the Internet of Things and cloud computing, will be harnessed to try to create more efficient, digitized and smart agricultural production.
这家动物园有各种各样的动物，包括熊、老虎、袋鼠和企鹅。

____.

5. Farmers will also receive training on how to better use information technology and their smart phones can become their new farm implements.
这位医生做了一场关于如何预防感冒的讲座。

____.

Passage 2

What Is the Difference Between Supply Chain and Logistics?

The terms "supply chain" and "logistics" are often used interchangeably within the transportation industry. They are, however, distinct areas, each involving specific processes, duties and responsibilities. The confusion in distinguishing between supply chain and logistics might stem from the fact that logistics is considered by many people to be a subcategory of supply chain management. The main difference between supply chain and logistics is that logistics is merely a specialized part of the entire supply chain process.

Generally, logistics focuses on the actual transportation and storage of goods. It deals with things such as inbound and outbound freight, reverse shipping, communications during transit, storage and warehousing. Logistics also deals with the delivery of goods and freight, coordination among third-party carriers, fleet management and other activities directly related to the actual transportation of goods from one point to another.

Supply chain management is the umbrella that covers all aspects of the sourcing and procurement of goods. Basically, supply chain management forms and manages the business-to-business links that allow for the ultimate sale of goods to consumers. Logistics, basically getting the freight from one place to the other, is a function that falls under the wide umbrella of supply chain management, but is only one part of the entire process.

The details and precise definitions for both the process of supply chain and logistics will vary from company to company and will overlap to a certain degree. Any person seeking to become involved in either supply chain management or logistics management within a company should ensure that the parameters of his or her responsibilities are clearly defined.

New Words

logistics	[lə'dʒɪstɪks]	*n.*	物流
interchangeably	[ˌɪntə'tʃeɪndʒəbli]	*adv.*	可互换地
distinct	[dɪ'stɪŋkt]	*adj.*	截然不同的
area	['eəriə]	*n.*	范畴
involve	[ɪn'vɒlv]	*v.*	包含
confusion	[kən'fju:ʒn]	*n.*	混淆
distinguish	[dɪ'stɪŋgwɪʃ]	*v.*	区分
subcategory	[sʌb'kætɪgəri]	*n.*	子范畴
merely	['mɪəli]	*adv.*	仅仅
specialized	['speʃəlaɪzd]	*adj.*	专门的
entire	[ɪn'taɪə(r)]	*adj.*	整个的
transit	['trænzɪt]	*n.*	运输
coordination	[kəʊˌɔ:dɪ'neɪʃn]	*n.*	协调
fleet	[fli:t]	*n.*	车队；船队
umbrella	[ʌm'brelə]	*n.*	综合体
cover	['kʌvə(r)]	*v.*	包括
aspect	['æspekt]	*n.*	方面
sourcing	['sɔ:sɪŋ]	*n.*	寻源采购；得到供货
procurement	[prə'kjʊəmənt]	*n.*	采购
form	[fɔ:m]	*v.*	形成；构形；组织
link	[lɪŋk]	*n.*	联系
ultimate	['ʌltɪmət]	*adj.*	最终的
precise	[prɪ'saɪs]	*adj.*	准确的
overlap	[ˌəʊvə'læp]	*v.*	部分重叠
parameter	[pə'ræmɪtə]	*n.*	范围；规范
define	[dɪ'faɪn]	*v.*	设定

Phrases & Expressions

stem from	根源是
focus on	集中于
deal with	涉及；处理
fall under	被归入
to a certain degree	在某种程度上

become involved in	参与
supply chain management	供应链管理
transportation industry	交通运输业
inbound and outbound freight	进出港货运
reverse shipping	逆向运输

Check your understanding

Decide whether the following statements are true or false according to the information given in the passage.

1. The terms "supply chain" and "logistics" are often used interchangeably within the construction industry.
2. The main difference between supply chain and logistics is that logistics is only a specialized part of the entire supply chain process.
3. Generally, logistics focuses on the actual transportation and sale of goods.
4. Supply chain management covers several aspects of the sourcing and procurement of goods.
5. The details and precise definitions for both the process of supply chain and logistics will overlap to a certain degree.

Task 4 Related Information

Understand the product information about the Norton AntiVirus Plus.

Norton AntiVirus Plus gets impressive scores in independent lab tests and our own hands-on tests and offers a wealth of useful features. However, it's expensive and it doesn't offer deals for multiple-computer households.

PROS

Excellent scores in independent lab tests and in our own tests.

Data Protector defends against ransomware.

Includes online backup, firewall, exploit protection, password manager and other bonus features.

CONS

Data Protector fared poorly in testing.

Expensive.

No multi-license pricing.

Task 5 Pop Quiz

Ⅰ. Working with words

Match the following English technical terms with their Chinese equivalents.

A. cloud computing
B. inbound and outbound freight
C. transportation industry
D. electronic payments
E. fleet management
F. reverse shipping
G. supply chain management
H. cold chain logistics
I. digitized agricultural production
J. online auctions
K. Internet banking
L. the Internet of Things

1. 冷链物流 (　　)	6. 物联网 (　　)
2. 网上拍卖 (　　)	7. 交通运输业 (　　)
3. 进出港货运 (　　)	8. 供应链管理 (　　)
4. 逆向运输 (　　)	9. 网上银行 (　　)
5. 电子支付 (　　)	10. 云计算 (　　)

Ⅱ. Multiple choice

1. E-commerce is a type of business model that does commercial transactions through the______.
   A. laptop　　B. smartphone　　C. Internet
2. ______ stands for business to consumer, which refers to transactions between online retailers and their customers.
   A. B2C　　B. B2B　　C. C2C
3. The e-commerce sector is already booming in rural areas thanks to the "______"model.
   A. Online　　B. Agriculture　　C. Internet Plus Agriculture
4. ______ such as processing, packaging, storage, preservation and cold chain logistics will be strengthened.
   A. Ways　　B. Support facilities　　C. Steps
5. Farmers will also receive training on how to better use information______.
   A. technology　　B. method　　C. measure
6. The terms "supply chain" and"______" are often used interchangeably within the transportation industry.
   A. cold chain　　B. logistics　　C. packaging
7. The main difference between supply chain and logistics is that logistics is merely a ______ part of the entire supply chain process.
   A. specialized　　B. digitalized　　C. modernized
8. ______ is the umbrella that covers all aspects of the sourcing and procurement of goods.

A. Storage management B. Business management C. Supply chain management

Ⅲ. Fill in the blanks with the proper choices given.

Antivirus Software

Antivirus software helps protect your computer against malware and cybercriminals. Antivirus software looks at data — web pages, files, software, applications — traveling over the network to your devices. It searches for ____(1)____, flagging suspicious behavior. It seeks to block or remove malware as quickly as possible.

Antivirus protection is essential, given the array of constantly emerging cyberthreats. ____(2)____, you could be at risk of picking up a virus or being targeted by other malicious software that can remain undetected and wreak havoc on your computer and mobile devices.

If you already have antivirus software, ____(3)____. But it might not be that simple. ____(4)____, it's important to stay current with the latest in antivirus protection.

If there's any crack in your cybersecurity defenses, cybercriminals likely will try to find a way in. Ensuring your antivirus software is up and running, and up-to-date, is a good place to start. However, ____(5)____, so it's a good idea to get protection from a comprehensive security solution.

A. you may believe you're all set

B. hackers, scammers, and identity thieves are constantly tweaking their methods

C. known threats and monitors the behavior of all programs

D. With new and savvier cyberthreats and viruses surfacing

E. If you don't have protective software installed

UNIT 3 Food Safety

Task 1 First Sight

1. Match the following pictures with the key words given.

A. purple cabbage　　B. sausage

C. instant noodles　　D. blueberry

(1)

(2)

(3)

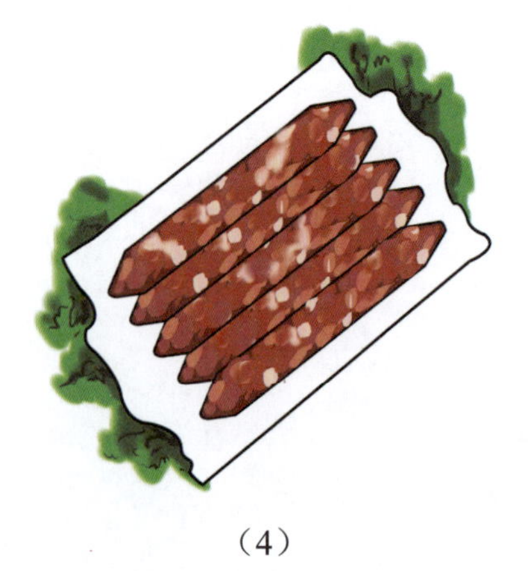
(4)

2. Answer the question by matching the following items.

How do you ensure food safety?

□1) use　　A. organic foods

□2) buy　　B. raw meats from other foods

□3) refrigerate — C. to the right temperature

□4) separate — D. baking soda to soak produce

□5) cook — E. foods promptly

Task 2 Better Acquaintance

Conversation

Are Processed Foods Unhealthy?

(Betty and Mary are talking about processed foods.)

Betty: Hi, Mary. You are eating crisps again. Yesterday you said that you wanted to lose weight.

Mary: Well, I just can't help eating them. They are so delicious.

Betty: That's why processed foods are very popular.

Mary: Processed foods are products that are heavily modified and contain a long list of ingredients. Am I right?

Betty: Yes, you are. They tend to be high in sugar, artificial ingredients, refined carbohydrates, and trans fats. They can have negative health effects.

Mary: It is said that many of them have minimal nutritional value.

Betty: That's true. Heating or drying foods can destroy certain vitamins and minerals.

Mary: In order to keep healthy, I will stop eating them.

Betty: Good. We can replace processed foods with whole foods, which include grains, nuts, seeds, lean meats, fruits, vegetables, and legumes.

Mary: That sounds great. I will do that.

New Words

process	[ˈprəʊses]	*v.*	加工
product	[ˈprɒdʌkt]	*n.*	产品
heavily	[ˈhevɪli]	*adv.*	在很大程度上
ingredient	[ɪnˈgriːdiənt]	*n.*	成分；配料
artificial	[ˌɑːtɪˈfɪʃl]	*adj.*	人工的
minimal	[ˈmɪnɪml]	*adj.*	极少的；最低的
nutritional	[njuˈtrɪʃənl]	*adj.*	营养的
value	[ˈvæljuː]	*n.*	价值

vitamin	[ˈvɪtəmɪn]	*n.*	维生素
mineral	[ˈmɪnərəl]	*n.*	矿物质
grain	[greɪn]	*n.*	谷物
seed	[si:d]	*n.*	籽
legume	[ˈlegju:m]	*n.*	豆科作物

Phrases & Expressions

tend to	往往会；常常就
lean meat	瘦肉
refined carbohydrate	精制碳水化合物
trans fat	反式脂肪
in order to	为了

Projects

Ⅰ. Substitution drills

1. Processed foods contain a lot of extra

糖.
油.
盐.
热量.
脂肪.

2. Some examples of processed foods include

即食餐.
糖果.
饼干.
炸薯条.
方便面.

3. Any food that's processed is usually

冷冻.
烘干.
罐装.
巴氏杀毒.
用盐腌制.

4. The processed meats that you should avoid are

培根.
香肠.
熟食肉.
火腿.
罐头肉.

5. Chemically processed foods

会对健康产生有害的影响.
含有较少维生素.
会导致体重增加.
几乎没有营养价值.
往往味道很好.

Ⅱ. The following are questions and answers about pesticides between two friends. Join the questions 1-8 to the answers A-H. Then act out some of the mini-dialogs with your partner.

1. How toxic are pesticides to humans? ________
2. Are pesticides toxic to humans? ________
3. Are there any pesticide residue limits? ________
4. What are pesticides? ________
5. Is there any food that is 100% pesticide-free? ________
6. Can pesticides affect human health? ________
7. Do animal products contain pesticide residues? ________
8. Do fruits and vegetables absorb pesticides? ________

A. Yes, pesticide residues are from the animal's own environmental exposure as well as through their feed.

B. No, almost no food is 100% pesticide-free. In fact, even organic farmers are still allowed to use certain pesticides of natural origin.

C. Yes, there are. WHO sets internationally-accepted maximum residue limits to protect food consumers from adverse effects of pesticides.

D. Yes, they can have both acute and chronic health effects, depending on the quantity and ways in which a person is exposed.

E. Yes. As they grow, they absorb pesticides and residues lingering on their skins.

F. They have the potential to harm the nervous system, the reproductive system and the endocrine system.

G. Pesticides are chemicals used to kill different kinds of pests.

H. Yes, they are.

Ⅲ. Fill in the blanks in the conversation by translating the Chinese into English.

Pesticide Residue Testing

（The manager needs to do pesticide residue testing. He is calling AC Produce Testing Center.）

Clerk: Good morning. AC Produce Testing Center. How can I help you?

Manager: Good morning. I need to do（1）________（农药残留检测）. Can you tell me something about it?

Clerk: Certainly. We offer you comprehensive testing service to（2）________（确认质量）of your vegetables and fruits. Our service also includes（3）________（检测前后分析）. In addition, we can test for a wide range of（4）________（可能的污染物）such as melamine, antibiotics and genetically modified organisms（GMOs）.

Manager: How do you carry out the testing?

Clerk: It is carried out at our laboratory, which is equipped with state-of-the-art analytical equipment. We use（5）________（不同的技术）such as gas and liquid chromatography for accurate results.

Manager: I see. Thank you very much.

Ⅳ. Remember the useful expressions concerning processed foods and pesticides.

1. They tend to be high in sugar, artificial ingredients, refined carbohydrates, and trans fats.
2. It is said that many of them have minimal nutritional value.
3. Heating or drying foods can destroy certain vitamins and minerals.
4. Almost no food is 100% pesticide-free.
5. Pesticides can have both acute and chronic health effects.
6. As fruits and vegetables grow, they absorb pesticides and residues lingering on their skins.
7. Pesticide residues are from the animal's own environmental exposure as well as through their feed.
8. It is carried out at our laboratory, which is equipped with state-of-the-art analytical equipment.

Task 3 Further Development

Passage 1

Pre-reading Task

Read the following statements and tick in the columns under "True", "False" or "Unsure".

Statements	True	False	Unsure
All the additives are good for human health.			
MSG is a harmless flavor enhancer.			
Artificial sweeteners may contain abnormal amino acids.			
Preservatives can't cause damage to the neurological system.			
Artificial ingredients have been linked to digestive issues.			

Four Food Additives to Avoid

It is undeniable that much of the food that we eat today is full of additives. Some of them are very dangerous and can pose serious health risks. Below you will find a list of the four most dangerous food additives that you need to avoid.

1. Monosodium Glutamate（MSG）

MSG is a harmful flavor enhancer. Many canned foods, soups, frozen meals, chips, and crackers contain MSG. Researchers believe that MSG can damage your body's cells and may even lead to cancer, learning disabilities, Parkinson's disease, and more.

2. Artificial Sweeteners

Many folks turn to artificial sweeteners because they are low calorie or no-calorie. Some researchers believe that artificial sweeteners, such as aspartame, may contain toxins, abnormal amino acids which may cause cancer. Some of the most common artificial sweeteners disguise themselves as aspartame, sucralose, saccharin, neotame, and acesulfame.

3. Preservatives

Two of the most common preservatives used to prolong shelf life of items are BHA and BHT. Research has suggested that these two preservatives can cause damage to the neurological system, and may even be related to cancer. Another common preservative is sodium nitrite, which is found in many deli meats and hot dogs. Sodium nitrites are believed to be linked to cancer.

4. Artificial Flavors and Colors

Anything that is artificial should probably not go into your body. Artificial ingredients have been linked to cognitive diseases, digestive issues, behavioral problems, and even cancer. In addition, a lot of young children have an allergy to these artificial colors and flavors, causing persistent and painful rashes.

New Words

undeniable	[ˌʌndiˈnaiəbl]	*adj.*	不可否认的
pose	[pəuz]	*v.*	造成
flavor	[ˈfleivə]	*n.*	味道
enhancer	[ɪnˈhɑːnsə(r)]	*n.*	增强剂
canned	[kænd]	*adj.*	罐装的
cell	[sel]	*n.*	细胞
damage	[ˈdæmɪdʒ]	*v.*	损坏
disability	[ˌdɪsəˈbɪləti]	*n.*	障碍
folk	[fəuk]	*n.*	人们
calorie	[ˈkæləri]	*n.*	卡路里（热量单位）
aspartame	[əˈspɑːteɪm]	*n.*	阿斯巴甜代糖
toxin	[ˈtɔksɪn]	*n.*	毒素
sucralose	['suːkrələʊs]	*n.*	三氯蔗糖
saccharin	[ˈsækərɪn]	*n.*	糖精
neotame	[niə'teɪm]	*n.*	纽甜
acesulfame	[əsɪsʌl'feɪm]	*n.*	安赛蜜
preservative	[prɪˈzɜːvətɪv]	*n.*	防腐剂
prolong	[prəˈlɒŋ]	*v.*	延长
cognitive	[ˈkɒgnətɪv]	*adj.*	认知的
digestive	[daɪˈdʒestɪv]	*adj.*	消化的
allergy	[ˈælədʒi]	*n.*	过敏反应
persistent	[pəˈsɪstənt]	*adj.*	持续的
rash	[ræʃ]	*n.*	皮疹

Phrases & Expressions

lead to	导致

disguise as 伪装成
shelf life 保质期
be related to 与……有关
be linked to 与……有关
monosodium glutamate 味精
artificial sweetener 人工甜味剂
abnormal amino acid 异常氨基酸
neurological system 神经系统
sodium nitrite 亚硝酸钠
deli meat 熟食肉

Projects

Ⅰ. Fill in the blanks with the proper words given, changing the form if necessary.

pose	prolong	calorie	addition
allergy	damage	persistent	lead

1. Eating too much sugar can __________ to health problems.
2. I have an __________ to peanuts.
3. Several vehicles were __________ in the crash.
4. Don't ignore a __________ cough.
5. Pesticides __________ a risk to bees.
6. The operation could __________ his life by two or three years.
7. One thin piece of bread has 90 __________ .
8. In __________ to processed foods, you have to avoid genetically modified foods.

Ⅱ. Translate the following sentences by imitating the examples given, paying attention to the underlined phrases or structures.

1. It is undeniable that much of the food that we eat today is full of additives.
他们将不可能赢得这场篮球赛。
____.

2. Below you will find a list of the four most dangerous food additives that you need to avoid.
汤姆喜欢我昨天买的书。
____.

3. Two of the most common preservatives used to prolong shelf life of items are BHA and BHT.
被用来切肉的刀有点钝了。
____.

4. Nitrates are believed to be linked to cancer.

他被公认为杰出的科学家。

____.

5. A lot of young children have an allergy to these artificial colors and flavors, causing persistent and painful rashes.

昨天下午发生了一起交通事故，造成七人死亡。

____.

Passage 2

The Pros and Cons of Genetically Modified Foods

Genetically modified foods have become a commonplace thing, even though few people understand just what "genetically modified" means. While there are some benefits that genetically modified foods may offer, there are also some risks and negative effects that these foods can cause.

When the term "genetically modified" is used to describe a food, it means that the genetic make-up of one of the ingredients in that food has been altered through genetic engineering.

There are several benefits that have been linked to genetically modified foods, including:

Resistance to disease: Genes can be modified to make crops more resilient when it comes to disease, especially those spread through insects.

Quality: Some genetically modified foods, particularly fruits and veggies, have a longer shelf life than natural products.

Taste: Some people claim that genetically modified foods have a better taste.

Nutritional content: Foods are often genetically modified in order to increase their nutritional content.

Although there are some benefits to genetically modified foods, there are some risks that have been associated with these foods. Some of these risks include:

Allergens and toxins: Some genetically modified foods may contain higher levels of allergens and toxins.

Antibiotic resistance: There may be an increased risk that people who eat those foods may be more resistant to antibiotics.

New diseases: Viruses and bacteria which are used in the process of modifying foods could cause the development of a new disease.

Nutritional content: Some genetically modified foods may actually lose nutritional content in the process of altering their genetic make-up.

New Words

modify	[ˈmɒdɪfaɪ]	*v.*	修改

commonplace	[ˈkɒmənpleɪs]	*adj.*	普遍的
negative	[ˈnegətɪv]	*adj.*	有害的
term	[tɜːm]	*n.*	词语
genetic	[dʒəˈnetɪk]	*adj.*	基因的
make-up	[meɪkʌp]	*n.*	组成成分
alter	[ˈɔːltə]	*v.*	改变
engineering	[ˌendʒɪˈnɪərɪŋ]	*n.*	工程
resistance	[rɪˈzɪstəns]	*n.*	抵抗力
gene	[dʒiːn]	*n.*	基因
resilient	[rɪˈzɪliənt]	*adj.*	有适应力的
spread	[spred]	*v.*	传播
insect	[ˈɪnsekt]	*n.*	昆虫
veggie	[ˈvedʒi]	*n.*	蔬菜
claim	[kleɪm]	*v.*	声称
associated	[əˈsəʊʃieɪtɪd]	*adj.*	有关联的
allergen	[ˈælədʒən]	*n.*	过敏原
antibiotic	[ˌæntibaɪˈɒtɪk]	*n.*	抗生素
virus	[ˈvaɪrəs]	*n.*	病毒
bacteria	[bækˈtɪəriə]	*n.*	细菌
process	[ˈprəʊses]	*n.*	过程

Phrases & Expressions

the pros and cons	事物的利与弊
even though	即使
when it comes to sth.	当涉及某事时
nutritional content	营养含量

Check your understanding

Answer the following questions according to the information given in the passage.

1. What does the term "genetically modified" mean?
2. Can modified genes make crops more resilient when it comes to disease?
3. Do genetically modified foods have a better taste according to some people?
4. Do genetically modified foods contain lower levels of allergens and toxins?
5. What could cause the development of a new disease?

Task 4 Related Information

Understand the product information about the bread improver.

A-800 Bread Improver

Features:

—a multi-purpose bread improver.

—increases the volume of bread.

—improves the bread tissue.

—improves dough fermentation stability.

Applications:

It is suitable for a variety of yeast dough, especially for the sweet bread.

Usage: Mix 0.3%—0.5% volume of improver with flour, water and other ingredients, then make the dough.

Packing: 1 kg×10/box

Task 5 Pop Quiz

Ⅰ. Working with words

Match the following English technical terms with their Chinese equivalents.

A. food additives　　B. pesticide residue

C. artificial sweetener　　D. preservative

E. genetically modified organisms　　F. organic foods

G. processed foods　　H. flavor enhancer

I. shelf life　　J. nutritional content

K. monosodium glutamate　　L. antibiotic resistance

1. 转基因生物（　　）	6. 加工食品（　　）
2. 增味剂（　　）	7. 保质期（　　）
3. 防腐剂（　　）	8. 食品添加剂（　　）
4. 农药残留（　　）	9. 营养含量（　　）
5. 有机食品（　　）	10. 人造甜味剂（　　）

Ⅱ. Multiple choice

1. ______ tend to be high in sugar, artificial ingredients, refined carbohydrates, and trans fats.

A. Organic foods　　B. Non-processed foods　　C. Processed foods

2. Heating or drying foods can destroy certain ______.

A. vitamins and minerals B. genes C. calories

3. ______ are chemicals used to kill different kinds of pests.

A. Additives B. Pesticides C. Bacteria

4. ______ can cause some risks and negative effects.

A. Organic fruits B. Organic vegetables C. Genetically modified foods

5. ______ is a harmful flavor enhancer.

A. GMO B. Preservative C. MSG

6. ______ can cause damage to the neurological system.

A. Artificial sweeteners B. Preservatives C. Proteins

7. You should avoid ______ for the sake of your health.

A. fruits B. vegetables C. deli meat

8. As fruits and vegetables grow, they absorb pesticides and ______ lingering on their skins .

A. residues B. viruses C. chemicals

Ⅲ. Fill in the blanks with the proper choices given.

The Importance of Food Safety

Access to sufficient amounts of safe and nutritious food is key to sustaining life and promoting good health. ___(1)___ can cause more than 200 different diseases—ranging from diarrhoea to cancers. Around the world, an estimated 600 million—almost 1 in 10 people—fall ill after eating contaminated food each year, ___(2)___ and the loss of 33 million healthy life years（DALYs）.

Food safety, nutrition, and food security are closely linked. Unsafe food creates a vicious cycle of disease and malnutrition, particularly affecting infants, young children, elderly, and the sick. ___(3)___, a safe food supply also supports national economies, trade, and tourism, stimulating sustainable development. The globalization of food trade, a growing world population, climate change and rapidly changing food systems ___(4)___. WHO aims to enhance at a global and country level the capacity to prevent, detect, and respond to public health threats ___(5)___.

A. In addition to contributing to food and nutrition security

B. have an impact on the safety of food

C. Unsafe food containing harmful bacteria, viruses, parasites, or chemical substances

D. associated with unsafe food

E. resulting in 420,000 deaths

UNIT 4 Food Processing

Task 1 First Sight

1. Match the following pictures with the key words given.

A. egg mixer   B. oven   C. flour mixer   D. bread slicer

(1)

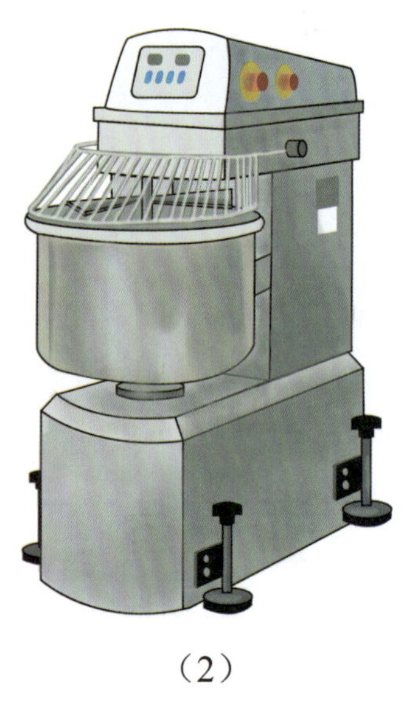

(2)

(3)

(4)

2. Answer the question by matching the following items.

What can baking machines do in everyday life?

□1) make	A. kitchens into bakeries
□2) turn	B. life easier
□3) replace	C. the way of baking food
□4) change	D. time and effort
□5) save	E. some common kitchen appliances

Task 2 Better Acquaintance

Conversation

How to Choose a Bread Machine

(Bob is curious about the bread machines on show. He is inquiring about them.)

Tom: Hello. How can I help you?

Bob: I'm interested in your bread machines. How do I choose the best one?

Tom: You should consider the type of machine, its baking time and the pan type.

Bob: What types do you have?

Tom: We have two types: automated and traditional.

Bob: Well, what's the difference between them?

Tom: The traditional machine requires a person to add the ingredients required. This type takes a bit more attention than the automated one.

Bob: How about the automated type?

Tom: It has different compartments to add the different ingredients into the machine, and it automatically mixes the ingredients. This machine takes less effort to make bread.

Bob: Sounds great. How long does it take to make bread?

Tom: 45 minutes to four hours. It largely depends on the type of bread and the machine itself.

Bob: Amazing machine!

New Words

automated	[ˈɔːtəmeɪtɪd]	*adj.*	自动化的
traditional	[trəˈdɪʃənl]	*adj.*	传统的

compartment	[kəmˈpɑːtmənt]	*n.*	分隔间；隔层
ingredient	[ɪnˈgriːdiənt]	*n.*	原料
automatically	[ˌɔːtəˈmætɪkli]	*adv.*	自动地

Phrases & Expressions

depend on（upon）　　取决于；依赖

Projects

Ⅰ. Substitution drills

1. Baking machines are available for a variety of foods, such as

面包.
蛋糕.
汉堡.
比萨饼.

2. High quality bread maker

使用方便.
容易清理.
容量大.
高效节能.

3. The baking ingredients include

面粉.
酵母.
鸡蛋.
糖.

4. Sugar can

增加食品的甜味.
改善食品的颜色和风味.
保持食品的水分.
有助于延长食品的保质期.

5. When purchasing a bread oven, you might consider

机器功能.
保修期限.
烘烤时间.
烤盘类型.

Ⅱ. The following are questions and answers about baking between two friends. Join the questions 1-8 to the answers A-H. Then act out some of the mini-dialogs with your partner.

1. What does the basic process of making bread involve? ________
2. Why make bread when I can buy it? ________
3. Can a bread machine make a cake? ________
4. Can we let the bread cool inside or outside the machine? ________
5. Do you read the food label when buying food? ________
6. Do you know how to read food labels? ________
7. How many types are food preservatives divided into? ________
8. Do you think salt is a food preservative? ________

A. Yes. Reading food labels can help us make healthy and safe food choices.

B. It involves mixing, kneading, rising and baking.

C. Because you will avoid all of the preservatives in the bread and be able to choose your ingredients entirely.

D. Yes. Salt is one of the most common food preservatives. It's often used in preserving canned fruits, processed meats as well as various types of canned or jarred vegetables.

E. Yes, of course. You can make a cake with your bread machine, and even jam and Japanese foods.

F. In addition to looking at the nutritional information on the label, we should also pay attention to the ingredients list, the claims made on the front of the label, and the amount of food contained in the package.

G. We recommend you remove it as soon as the process ends, as leaving it inside will cause the top of the bread to soften.

H. Two types. Natural preservatives and artificial preservatives.

Ⅲ. Fill in the blanks in the conversation by translating the Chinese into English.

How to Make a Berry Pie

(Amy and her friend Tina are talking about how to make a berry pie.)

Amy: I love berries. Here are raspberries I (1) ______________ (采摘) in my garden just now.

Tina: So fresh! Do you know how to make a berry pie?

Amy: Yeah. I also love pies. Let's get started.

Tina: OK. What are needed to make a berry pie?

Amy: First we need to gather all the ingredients that are needed, and then make the pie filling.

Tina: What do the ingredients include?

Amy:（2）________（黄油）, sugar, water, cornstarch,（3）________（柠檬汁）and cinnamon.

Tina: Do we need to add them to the berries and mix them all together?

Amy: You are right. Then we can get delicious berry pie filling.

Tina: Wow! It smells great! What's the next step?

Amy: Mix（4）____________（面粉）with butter, water and sugar. We can get the berry pie crust.

Tina: Do we need to put the filling on the crust next?

Amy: Yes. Then place the pie in（5）___________（烤箱）and soon we'll get a delicious berry pie.

Tina: Oh, making a berry pie is so much fun!

Ⅳ. Remember the useful expressions concerning baking machines.

1. Do you have a bread machine?
2. How does the machine work?
3. What does a bread machine do?
4. How do I choose the best bread machine?
5. What kind of baking machine do you have?
6. How long does it take to make bread?
7. What are the features of the oven?
8. What does the basic process of making bread involve?

Task 3 Further Development

Passage 1

Pre-reading Task

Read the following statements and tick in the columns under "True", "False" or "Unsure".

Statements	True	False	Unsure
The farmers behind commodity meat often meet the people they supply meat to.			
Meat processing plants don't slaughter animals, only process or store meat.			
Most farmers behind commodity meat are separated by a long chain of feedlots, slaughterhouses and distributors.			
Commodity meat is quite cheaper than meats produced by small producers because of the large amount of the production.			
Some people believe that commodity meat is bad for the environment.			

Commodity Meat

The term "commodity meat" is sometimes used to describe meat which is produced on the industrial scale and sold at relatively low cost. In contrast with commodity meat, consumers may choose to purchase products from smaller producers which have an emphasis on humane and/or sustainable production, but they must be prepared to pay a premium for those products.

Animals raised for commodity meat and animal products are classically raised in Confined Animal Feeding Operations, known as CAFOs. CAFOs are designed to be as space efficient as possible, which means that large numbers of animals may be packed into a small area. The animals are typically fed with grains purchased with the assistance of government subsidies, and they may lack access to a varied diet or the outdoors. CAFOs are also infamous for generating huge volumes of manure and other waste materials, which can create a polluting issue.

Commodity meat is vastly cheaper than meats produced by small producers both because of the sheer volume of the production, and because farmers benefit from government subsidies. The government may also choose to purchase commodity meat and sell it at a discount to institutions like schools, hospitals, and prisons. People who would prefer to see more humane farming are often frustrated by the low price of commodity meat, which makes it harder for consumers to switch meat sources. Unless you are told otherwise, you should assume that all meat you eat in restaurants or buy in stores is commodity meat.

New Words

commodity	[kəˈmɒdəti]	*n.*	商品
emphasis	[ˈemfəsɪs]	*n.*	强调
humane	[hjuːˈmeɪn]	*adj.*	人道的
sustainable	[səˈsteɪnəbl]	*adj.*	可持续的
premium	[ˈpriːmiəm]	*n.*	额外费用
classically	[ˈklæsɪkli]	*adv.*	最常见地
confined	[kənˈfaɪnd]	*adj.*	狭窄而围起来的
subsidy	[ˈsʌbsədi]	*n.*	补贴
generate	[ˈdʒenəreɪt]	*v.*	产生
manure	[məˈnjʊə(r)]	*n.*	粪肥

vastly	[ˈvɑ:stli]	*adv.*	非常
sheer	[ʃɪə(r)]	*adj.*	完全的
institution	[ˌɪnstɪˈtju:ʃn]	*n.*	机构
frustrate	[frʌˈstreɪt]	*v.*	使懊恼
switch	[swɪtʃ]	*v.*	转变
assume	[əˈsju:m]	*v.*	假设
otherwise	[ˈʌðəwaɪz]	*adv.*	另外

Phrases & Expressions

in contrast with	与……相比
benefit from	受益于

Projects

Ⅰ. Fill in the blanks with the proper words given, changing the form if necessary.

frustrate	switch	emphasis	sustainable
assume	confined	generate	subsidy

1. He placed great ________ on his children's education.
2. He's decided to ________ the meeting from Monday to Thursday.
3. The government is planning to increase ________ for farmers.
4. What ________ him is that there's too little money to spend on the project.
5. It is generally ________ that stress is caused by too much work.
6. It is cruel to keep animals in ________ spaces.
7. The proposal ________ a lot of interest.
8. A large international meeting was held with the aim of promoting ________ development in all countries.

Ⅱ. Translate the following sentences by imitating the examples given, paying attention to the underlined phrases or structures.

1. The term "commodity meat" is sometimes used to describe meat which is produced <u>on the industrial scale</u>.

   这个地区将<u>大规模地</u>种植农作物。

   ____.

2. <u>In contrast with</u> commodity meat, consumers may choose to purchase products from smaller

producers.

与他们的建议相比，我们的似乎更切实可行。

____.

3. They may lack access to a varied diet or the outdoors.

学生必须有机会使用好的资源。

____.

4. Commodity meat is vastly cheaper than meats produced by small producers.

昨天被邀请出席生日宴会的大多数人都是我的同学。

____.

5. Farmers benefit from government subsidies.

我觉得，从她的睿智中我获益良多。

____.

Passage 2

Fermented Milk

Fermented milk, also known as cultured milk, is a type of dairy food which is made by adding lactic acid bacteria, mold, or yeast to milk. The specific chemical reaction and product that results from fermentation depends upon the type of bacteria used and the process by which it is combined with the milk. It is commonly used to create dairy products such as yogurt, cheese, sour cream, and buttermilk. Fermented milk was first made in order to increase the shelf life of dairy products. It can also make milk easier to digest and enhance the flavor and texture of dairy foods.

The form and flavor of a fermented dairy food depends upon the type of milk product and the way in which the bacteria or mold are introduced. For example, yogurt and cheese are made with milk, while sour cream is started with light cream. Most milk products that are made from fermentation have at least one form of lactic acid bacteria. Some also have a specific type of mold or yeast in addition to the bacteria.

Fermented milk products can be beneficial to health, as the process often makes them more easily digestible for many individuals. In addition to products with beneficial live cultures like yogurt, bacteria such as acidophilus can be added to milk in order to make it more easily managed by those with lactose intolerance. There has also been evidence that certain fermented dairy foods can lower cholesterol and help ease diarrhea and other symptoms of inflammatory bowel diseases.

New Words

ferment	[fəˈment]	*v.*	（使）发酵
cultured	[ˈkʌltʃəd]	*adj.*	培养的

lactic	[ˈlæktɪk]	*adj.*	乳的
acid	[ˈæsɪd]	*n.*	酸
bacteria	[bækˈtɪəriə]	*n.*	细菌
mold	[məʊld]	*n.*	霉菌
yeast	[jiːst]	*n.*	酵母；酵母菌
fermentation	[ˌfɜːmenˈteɪʃn]	*n.*	发酵（作用）
combine	[kəmˈbaɪn]	*v.*	（使）结合，组合
buttermilk	[ˈbʌtəmɪlk]	*n.*	脱脂乳
digest	[daɪˈdʒest]	*v.*	消化
enhance	[ɪnˈhɑːns]	*v.*	提高；增强
texture	[ˈtekstʃə(r)]	*n.*	质地；口感
digestible	[daɪˈdʒestəbl]	*adj.*	易消化的；口感好的
acidophilus	[ˌæsiˈdɒfələs]	*n.*	嗜酸菌；乳酸杆菌
lactose	[ˈlæktəʊs]	*n.*	乳糖
intolerance	[ɪnˈtɒlərəns]	*n.*	不耐受
cholesterol	[kəˈlestərɒl]	*n.*	胆固醇
diarrhea	[ˌdaɪəˈriːə]	*n.*	腹泻
symptom	[ˈsɪmptəm]	*n.*	症状
inflammatory	[ɪnˈflæmətri]	*adj.*	发炎的；炎性的
bowel	[ˈbaʊəl]	*n.*	肠

Phrases & Expressions

result from	由……产生；由……引起
combine…with…	与……结合
shelf life	货架期，保质期
sour cream	酸奶油（烹饪用）
light cream	淡奶油
at least	至少
lactic acid bacteria	乳酸菌
in addition to	除……之外

Check your understanding

Answers the following questions according to the information given in the passage.

1. What is fermented milk?

2. What's the original purpose of making fermented milk?

3. What does the form and flavor of a fermented dairy food mainly depend upon?

4. Why can fermented milk products be beneficial to health?

5. Why can acidophilus be added to milk?

Task 4 Related Information

Understand the product information about the rotary oven.

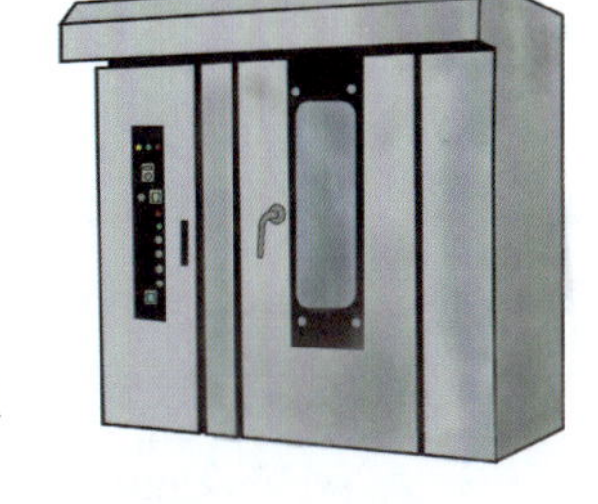

Product Description:

Used for a large number of fast food baking, apply to food factory, large chain stores and other large quantities of baking.

Available for food: moon cake, toast, all kinds of bread, hamburger, all kinds of snacks, biscuits, meat, etc.

Rotary Oven					
Model	Type	Specification	Dimension（mm）	Voltage	Power
HG-16D	Electric Rotary Oven	16 Trays	1,400×2,235×2,420	380 V/3 N	32 kW
HG-32D		32 Trays	1,660×2,600×2,420	380 V/3 N	46 kW
HG-64D		64 Trays	2,008×2,885×2,420	380 V/3 N	72 kW
HG-16Q	Gas Rotary Oven	16 Trays	1,400×2,235×2,420	380 V/3 N	3.0 kW
HG-32Q		32 Trays	1,660×2,600×2,420	380 V/3 N	3.0 kW
HG-64Q		64 Trays	2,008×2,885×2,420	380 V/3 N	3.0 kW
HG-16Y	Diesel Rotary Oven	16 Trays	1,400×2,235×2,420	380 V/3 N	3.0 kW
HG-32Y		32 Trays	1,660×2,600×2,420	380 V/3 N	3.0 kW
HG-64Y		64 Trays	2,008×2,885×2,420	380 V/3 N	3.0 kW

Task 5 Pop Quiz

Ⅰ. Working with words

Match the following English technical terms with their Chinese equivalents.

A. oven

B. ingredients required

C. bread slicer

D. lactose intolerance

E. egg mixer

F. dairy products

G. pan type

H. fermented milk

I. flour mixer　　　　J. processed meat

K. lactic acid bacteria　　　　L. canned fruit

1. 乳酸菌（　　）	6. 烤箱（　　）
2. 打蛋机（　　）	7. 水果罐头（　　）
3. 发酵乳（　　）	8. 乳糖不耐受（　　）
4. 加工肉（　　）	9. 所需原料（　　）
5. 奶制品（　　）	10. 面包切片机（　　）

Ⅱ. Multiple choice

1. When you read the nutritional label, you can get the basic information about ______.

A. the nutritional content of the food

B. the ingredients

C. storage conditions

2. Yogurt and cheese are made with ______.

A. wheat flour　　B. milk　　C. bean flour

3. When you decide to buy a bread machine, it's important to consider ______.

A. the type of machine　　B. its baking time　　C. Both A and B

4. Food additives can not be found in ______.

A. store-bought bread　　B. fresh milk　　C. soy sauce

5. Color additives may be used ______.

A. to make foods feel softer　　B. to make foods taste better　　C. to make foods look better

6. ______ can be used to measure the ingredients you need.

A. Measuring cups　　B. Measuring spoons　　C. Both A and B

7. The recipe gives you all of the ingredients you need as well as ______ on how to cook something.

A. the detailed instructions　　B. the experience　　C. the choice of appliances

8. When you take the cookies out of the oven, you'd better put them on a ______ to cool off for a bit.

A. baking sheet　　B. cooling rack　　C. baking pan

Ⅲ. Fill in the blanks with the proper choices given.

A lot of interesting and important information can be found on a food label, ___(1)___ or aren't sure about how they should interpret the information on food labels. Learning to read food labels can be an important step in improving nutrition, ___(2)___, as reading labels can help savvy shoppers pick out the best nutritional and financial choices in the store. In addition to looking at the nutritional information on the label, shoppers should also pay attention to the ingredients list, ___(3)___, and the amount of food contained in the package.

When people think about reading food labels, ___(4)___. Nutritional labeling is designed to provide people with basic information about the nutritional content of the food, so that they can make sound choices. The top of a nutritional level indicates the size of a serving, and the number of servings in a package: it's important to pay attention to this, because a serving size might be smaller than expected, especially with high-calorie foods.

The nutritional label also states the number of calories in a serving. As a general rule, the label on a nutritionally sound item should indicate that ___(5)___, and low in fats, carbohydrates, and sugars.

A. they often think of the nutritional label

B. the food is high in fiber, protein, and vitamins

C. the claims made on the front of the label

D. and it can also save people money

E. but some people don't read food labels

UNIT 5 Ecotourism

Task 1 First Sight

1. Match the following pictures with the key words given.

A. garbage classification B. landfill

C. ecological environment D. garbage pollution

(1)

(2)

(3)

(4)

2. Answer the question by matching the following items.

What should you do with your garbage in everyday life?

□1）recycle	A. toxic waste carefully
□2）convert	B. garbage into dry and wet
□3）bury	C. waste like metal, glass and plastic
□4）dispose of	D. food waste to compost
□5）classify	E. garbage in a landfill

Task 2 Better Acquaintance

Conversation

Garbage Pollution on the Beach

（Paul is not happy with his vacation. He is talking about it with Amy.）

Amy: Hi, Paul. Where did you go for your vacation?

Paul: I went to the beach with some friends.

Amy: Did you have a good time?

Paul: Actually not. Trash! Trash everywhere! Everyone loves a clean beach, but too many people also love to trash the beach.

Amy: Oh, sorry to hear that. Why is that the case?

Paul: Maybe it's not convenient to find a garbage can.

Amy: What's so bad about garbage on the beach?

Paul: Broken glass, fishing line and other waste can harm sea life and tourists.

Amy: Can it also bring some changes to the ecosystem?

Paul: Definitely. Beach garbage affects many living and non-living organisms.

Amy: Well, what can we do about it now?

Paul: The most effective way is to target waste on the beach, and make sure it never reaches the ocean.

Amy: OK. Let's do our part to keep it clean. First bring reusable bags instead of plastic ones when leaving the house.

Paul: Great. We're working towards cleaner coasts and healthier oceans.

New Words

garbage	[ˈgɑːbɪdʒ]	*n.*	垃圾；废物
trash	[træʃ]	*n./v.*	垃圾；弄脏；弄乱
ecosystem	[ˈiːkəʊsɪstəm]	*n.*	生态系统

organism	[ˈɔːɡənɪzəm]	*n.*	有机体；生物
reusable	[ˌriːˈjuːzəbl]	*adj.*	可重复使用的
coast	[kəʊst]	*n.*	海岸；海滨

Projects

Ⅰ. Substitution drills

1. Ecotourism can

   保护当地生物多样性.
   促进资源的可持续利用.
   支持当地经济.
   提高资源保护的认识.

2. Piles of garbage

   占用土地.
   污染生活环境.
   危害人的健康.
   影响农作物生长.

3. Garbage can be classified into

   干垃圾.
   湿垃圾.
   可回收物.
   有害垃圾.

4. Waste can be managed by

   回收利用.
   填埋.
   制成肥料.
   焚烧.

5. Recycling waste can

   变废弃物为新材料.
   减少污染.
   节省材料.
   降低能源消耗.

Ⅱ. The following are questions and answers about ecotourism between two friends. Join the questions 1-8 to the answers A-H. Then act out some of the mini-dialogs with your partner.

1. Do you have any plans for your weekend? ________
2. Where do you like to go in summer? ________
3. What can you do on the ranch in summer? ________
4. What do you think of ecotourism? ________
5. What can tourists learn at an ecotourism site? ________
6. Can local people benefit from ecotourism? ________
7. Can you say something about rural tourism? ________
8. What's the difference between natural tourism and ecotourism? ________

A. Yes. I love berries. I'm going to visit a farm and pick some.

B. There are so many activities that we can do on the ranch. We can go hiking in the woods. We can also see the beautiful animals there.

C. In summer, we like to go visit the ranch. A lot of ranches have big open fields of green grass. We can enjoy the stunning natural beauty.

D. They can learn about the ecosystems, traditions and cultures of their destination. They can also increase their environmental awareness.

E. Ecotourism is responsible travel to natural areas that conserves the environment and improves the well-being of local people.

F. It is a form of ecotourism that attracts tourists to the rural culture.

G. Yes. Ecotourism can provide jobs and income for local people.

H. They have different purposes. For example, both of them may involve hiking and camping, but natural tourism is for the entertainment and recreation, and ecotourism is for learning about the area or making an improvement to it.

Ⅲ. Fill in the blanks in the conversation by translating the Chinese into English.

Protecting Environment

(Amy and Chris are on vacation. They are talking about environmental problems when they see so much garbage everywhere.)

Amy: Environmental pollution is increasingly serious now.

Chris: It's really worrying. We throw away so much garbage every day.

Amy: Garbage pollution brings about a series of (1) ________________ (生态和环境问题). Forests are gone, grassland is destroyed, rivers and lakes are polluted and wild animals are (2) ________________ (濒临灭绝). It is horrible.

Chris: So, what should we do while enjoying (3) ________________ (自然美景)?

Amy: It's very simple. Do not litter, and use as few （4）_______________（塑料袋）as possible.

Chris: Is there anything else we can do for it?

Amy: Yes. Waste must be classified before being put into （5）_______________（垃圾箱）.

Chris: Where does all that trash go?

Amy: Recycling, landfilling, composting and incinerating. Each method has its strengths and weaknesses.

Chris: Does recycling actually help?

Amy: Yes. Recycling is the process of turning waste into new materials. For example, used paper can be turned into cardboard.

Chris: Terrific! Let's work together and start with garbage classification.

Ⅳ. Remember the useful expressions concerning ecotourism.

1. Where did you go for your last vacation?
2. What's your plan for this weekend?
3. Have you ever been to an ecotourism site?
4. What special places would you like to see?
5. Will you join a package tour?
6. What can we see there?
7. How did you go there?
8. Did you have a good time?

Task 3 Further Development

Passage 1

Pre-reading Task

Read the following statements and tick in the columns under "True", "False" or "Unsure".

Statements	True	False	Unsure
Ecotourists can gain knowledge of ecosystems, biology and geology of specific natural locations.			
Many of the world's most beautiful natural sites exist in developed countries.			
All travel organizations that market themselves as ecotourism programs are actually environmentally friendly.			
Personal responsibility is a large part of the ecotourism experience.			
Tourists can enjoy the recreational aspects of a trip while having a low impact on the environment.			

What Is Ecotourism?

Ecotourism is a form of tourism which places a heavy emphasis on appreciation and protection of the natural environment, with ecotourists traveling to regions of ecological interest around the world. This form of tourism is also sometimes called ecological tourism, nature travel, or responsible tourism. Like other forms of tourism, ecotourism touches on some very complex environmental, social, and ethical issues, and a number of professional organizations have banded together to create a firm definition for ecotourism so that standards can be established for ecotourism programs.

In order to qualify as ecotourism, several criteria must be met. The most important criterion is, in the eyes of many people, minimal environmental impact, as people do not want to damage the natural environment while they are trying to appreciate it. Ecotourism also typically includes an educational aspect, with visitors learning about the environments they visit, and there is a heavy emphasis on conservation. In some cases, people may even participate in a service program on an ecotourist trip, doing something to actively benefit the environment while enjoying it.

Ecotourism is especially popular in Africa, South America, and Asia, where stretches of largely untouched land still exist in some regions. Tourists can travel to various locations by animals, boats, or on foot, and while on location, they are typically encouraged to camp or use basic facilities provided by the tourist companies. Companies which cater to ecotourists typically minimize luxuries, with the understanding that luxury often has a negative environmental impact. Once on site, the tourists may participate in guided trips, visit interesting sites in the area, or interact with native people to learn more about their culture.

New Words

appreciation	[əˌpriːʃiˈeɪʃn]	*n.*	欣赏
ecological	[ˌiːkəˈlɒdʒɪkl]	*adj.*	生态的
complex	[ˈkɒmpleks]	*adj.*	复杂的
ethical	[ˈeθɪkl]	*adj.*	道德的
definition	[ˌdefɪˈnɪʃn]	*n.*	定义
standard	[ˈstændəd]	*n.*	标准
establish	[ɪˈstæblɪʃ]	*v.*	建立
qualify	[ˈkwɒlɪfaɪ]	*v.*	符合；使合格

criterion	[kraɪˈtɪəriən]	*n.*	标准
minimal	[ˈmɪnɪməl]	*adj.*	最小的
aspect	[ˈæspekt]	*n.*	方面
conservation	[ˌkɒnsəˈveɪʃn]	*n.*	保护
participate	[pɑːˈtɪsɪpeɪt]	*v.*	参加
facility	[fəˈsɪləti]	*n.*	设施
stretch	[stretʃ]	*n.*	一片
luxury	[ˈlʌkʃəri]	*n.*	奢侈的享受；奢侈品

Phrases & Expressions

place emphasis on	重视
band together	联合
in the eyes of	在……看来
participate in	参与
cater to	迎合
interact with	与……互动

Projects

Ⅰ. Fill in the blanks with the proper words given, changing the form if necessary.

ecological	appreciation	ethical	definition
establish	participate	qualify	criterion

1. There is no general agreement on a standard ________ of intelligence.
2. Practice is the sole ________ for testing truth.
3. The school ________ a successful relationship with the local community.
4. This training course will ________ you for a better job.
5. We encourage students to ________ fully in the running of the college.
6. Transplantation of organs from living donors raises ________ issues.
7. Large dams have harmed Siberia's delicate ________ balance.
8. She shows little ________ of good music.

Ⅱ. Translate the following sentences by using the phrases or expressions given in the brackets.

1. 很多学校开始注重使用电脑教学。（place emphasis on）

2. 在我父母的心目中，我永远是个孩子。（in the eyes of）
3. 你可以去迎合他人，但是不要忘记你是谁。（cater to）
4. 银行积极鼓动人们贷款。（encourage sb. to do sth.）
5. 我喜欢和人们互动，我的沟通能力很好。（interact with）

Passage 2

Living Green

Living green means having a lifestyle that is environmentally conscious. It means being earth-friendly or environmentally friendly, rather than doing things that are harmful to our world. In general, living green can be accomplished through doing what is known as "the 3 Rs": recycling, reusing and reducing.

Reducing waste helps lower the amount of garbage in landfills. Garbage piled up causes pollution; it's difficult to dispose of cleanly and some of it ends up in the oceans. Some groups focused on green living have protested the amount of packaging that manufacturers use in making products, such as having an item in a box with plastic wrap over it. Many companies today have new packaging designs that are more environmentally friendly, resulting in less waste.

One of the most important ways of living green is to reduce carbon emissions from vehicles. Emissions from cars are a strong threat to sustainable living. It's a known fact that if we continue to pollute the earth, it will no longer be a sustainable environment for future generations.

Reusing items helps keep them from piling up in the landfill. Donating still usable, but unwanted clothing and household goods to people or organizations allows the items to have a second life rather than having to be processed as garbage. Reusing stained or ripped clothing as cleaning rags is another way of living green. In addition to being reused, cloth rags cut down or eliminate the amount of paper towels needed in a household. Unless they're made from recycled materials, paper towels aren't considered environmentally friendly as the pulp they consist of comes from natural resources including trees.

New Words

conscious	[ˈkɒnʃəs]	*adj.*	有意识的
accomplish	[əˈkʌmplɪʃ]	*v.*	完成
recycle	[ˌriːˈsaɪkl]	*v.*	回收利用
landfill	[ˈlændfɪl]	*n.*	废物填埋地
protest	[prəˈtest]	*v.*	抗议
packaging	[ˈpækɪdʒɪŋ]	*n.*	包装
manufacturer	[ˌmænjuˈfæktʃərə(r)]	*n.*	生产商
carbon	[ˈkɑːbən]	*n.*	碳

emission	[iˈmɪʃn]	*n.*	排放
vehicle	[ˈviːəkl]	*n.*	交通工具
threat	[θret]	*n.*	威胁
sustainable	[səˈsteɪnəbl]	*adj.*	可持续的
donate	[dəʊˈneɪt]	*v.*	捐赠
stained	[steɪnd]	*adj.*	沾有污渍的
ripped	[rɪpt]	*adj.*	撕破的
rag	[ræg]	*n.*	破布
pulp	[pʌlp]	*n.*	纸浆

Phrases & Expressions

be harmful to	对……有害
in general	总的来说
pile up	（使）成堆
dispose of	处理
focus on	关注
result in	导致
no longer	不再
cut down	削减

Check your understanding

Choose the correct answers according to the information given in the passage.

1. Living green means being earth-friendly or environmentally friendly, rather than doing things like ________.

   A. planting trees

   B. throwing out garbage everywhere

   C. throwing garbage into dustbins

2. Living green can be accomplished through doing what is known as "the 3 Rs": recycling, reusing and ________.

   A. replacing　　B. renewing　　C. reducing

3. Many companies today have new packaging designs that are more environmentally friendly, resulting in ________.

   A. less waste　　B. more waste　　C. more pollution

4. Waste consisting of cans, ________, metal, glass, and paper can be recycled.

   A. batteries B. unwanted clothing C. discarded medicine

5. Unless they're made from recycled materials, paper towels aren't considered environmentally friendly because ________.

   A. it pollutes the environment

   B. it affects crops growth

   C. it causes more trees to be cut down

Task 4 Related Information

Read the leaflet about the U-Pick farm and judge whether the explanation of their selling point is reasonable or not.

Summer Activities with the Kids: Go Berry, Apple, or Peach Picking!

Looking for some fun activities to do with your kids this summer?

Today's idea: Go Berry, Apple, or Peach Picking!

Through the years and beyond all the less-than-fun parts, our family found these fruit picking experiences to be wonderful because:

> The kids got to learn more about where our food comes from.
>
> The kids got to learn about the hard work it takes to put food on the table.
>
> We got to take home loads of delicious, nutritious fruits to eat in a variety of fun ways.
>
> It saved all kinds of money as we either had to pay very small prices per pound of fruits, or we were encouraged to take all we wanted for free!

Task 5 Pop Quiz

I. Working with words

Match the following English technical terms with their Chinese equivalents.

A. ecosystem B. ecotourism

C. garbage classification D. recyclable waste

E. garbage pollution　　F. environmental pollution

G. dry waste　　H. wet waste

I. toxic waste　　J. landfill

K. incineration　　L. compost

1. 生态旅游（　）	6. 可回收垃圾（　）
2. 生态系统（　）	7. 垃圾填埋场（　）
3. 垃圾污染（　）	8. 湿垃圾（　）
4. 垃圾分类（　）	9. 有害垃圾（　）
5. 焚毁（　）	10. 堆肥（　）

Ⅱ. Multiple choice

1. Ecotourism is a form of tourism which places a heavy emphasis on ________.

A. appreciation and protection of the natural environment

B. exploiting the natural resources heavily

C. altering beautiful natural landscapes

2. Garbage is primarily classified into ________, dry waste and wet waste.

A. recyclable waste　　B. toxic waste　　C. Both A and B

3. Toxic waste includes batteries, ________, dried paint, old bulbs, and dried shoe polish.

A. glass　　B. pesticide　　C. plastic

4. The wet or organic waste should be put in ________ and the resulting compost could be used as manure in the garden.

A. a compost pit　　B. landfill　　C. garbage can

5. Tourists are typically forbidden to ________ when visiting ecotourism sites.

A. camp　　B. use basic facilities　　C. do harm to the environment

6. Nobody wants to live next to a ________.

A. park　　B. landfill　　C. lake

7. While admiring the natural view and scenery, you must try to ________.

A. trample the native flora

B. buy souvenirs made from wild animal products

C. avoid using soaps in fresh waterways

8. The choice of travel will depend on traveler's preference, ________.

A. budget　　B. length of stay　　C. Both A and B

Ⅲ. Fill in the blanks with the proper choices given.

Best Destinations for a Vacation in Europe

Traveling to Europe is a dream of many, but with so many different experiences to choose from, it may be difficult to choose the best vacation in Europe. ____(1)____, and culture found across the region, so it is important to narrow down choices for a vacation in Europe to ___(2)___. Some of the best choices for a vacation in Europe include ancient and grand cities, unspoiled countryside, and romantic seaside locations.

Europe is home to more than a dozen spectacular cities boasting fascinating sights. In London, ___(3)___, taking marvelous walking tours, and enjoying some of the best live theater performances in the world. Paris is a famously dramatic city known for magnificent vistas, indulgent cafes featuring famous French cuisine, and fabulous shopping. For romantics, ___(4)___, with its wandering streets, incredible museums, and numerous sculptures and fountains by the greatest artistic masters in history. For fewer tourists and better prices while on a city vacation in Europe, consider visiting Prague or Amsterdam, ___(5)___.

A. visitors can spend hours touring historic palaces

B. It could take a lifetime to explore all of the beauty, history

C. few cities can beat the ancient allure of Rome

D. both known for their unique architecture and fascinating history

E. match personal interests and style

UNIT 6 Hotel Business

Task 1 First Sight

1. Match the following pictures with the key words and phrases given.

A. reception desk B. check-in form C. key card D. a suite

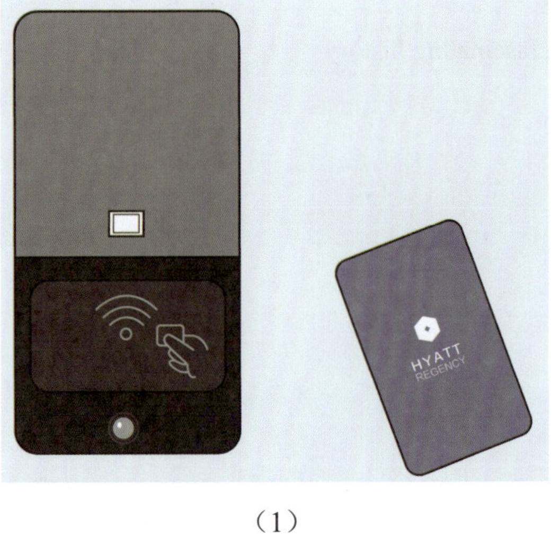

(1)

(2)

(3)

酒 店 住 宿 登 记 表 NO0009551

序号: _楼_房

姓　名	性别	年龄	工 作 单 位	职业
户口地址				
身份证或其他有证件名称				
身份证或其他有证件名称				

(4)

2. Answer the question by matching the following items.

What factors do you usually consider when choosing a hotel?

□1） free	A. price
□2） reasonable	B. environment
□3） various	C. Wi-Fi
□4） convenient	D. traffic
□5） comfortable	E. facilities

Task 2 Better Acquaintance

Conversation

At the Front Desk

（A guest comes to the Holiday Hotel. He is checking in at the reception desk.）

A: Good morning, sir! Welcome to Holiday Hotel. How may I help you?

B: I'd like to check in, please.

A: Do you have a reservation?

B: Yes. Under Scott.

A: OK, let me check. A single room with bath for two nights. Is that right?

B: That's right.

A: Well, your room is on the second floor. Would you please reconfirm all the details and sign your name here?

B: OK. By the way, is breakfast included in the rate?

A: Yes. Here is your key card. You can take the elevator on your left.

B: Thank you very much.

A: It's my pleasure. Please enjoy your stay.

B: Thank you .

New Words

reservation	[rezə'veɪʃ(ə)n]	*n.*	预订
single	['sɪŋg(ə)l]	*adj.*	单一的
confirm	[kən'fɜːm]	*v.*	确认
detail	['diːteɪl]	*n.*	细节
sign	[saɪn]	*v.*	签字

Phrases & Expressions

check in	入住
have a reservation	有预订
on the second floor	在二楼
on one's left	在某人左手边

Projects

Ⅰ. Substitution drills

1. I'd like to
   - 了解洗衣服务.
   - 了解一下双人间的房价.
   - 买一张去北京的单程车票.
   - 知道附近有什么好玩的地方.
   - 知道早餐时间.

2. What kind of room do you want?
   - 双人间.
   - 三人间.
   - 套房.
   - 标准间.
   - 豪华间.

3. I'd like to book a single room with
   - 海景.
   - 阳台.
   - 网络.
   - 城市景观.
   - 空调.

4. Your room is
   - 在五楼.
   - 在经理办公室对面.
   - 在会议室旁边.
   - 左数第三个.
   - 在地下室.

5. Would you please

填写表格?
出示您的身份证?
稍等一下?
扫描这个二维码?
确认这些信息?

Ⅱ. The following are questions and answers about hotel services between a customer and a hotel clerk. Join the questions 1-8 to the answers A-H. Then act out some of the mini-dialogs with your partner.

1. How many pieces of luggage do you have? ________
2. Good morning! Friendship Hotel room reservation. May I help you? ________
3. Could you call a taxi for us ? We've just checked out. ________
4. Could I settle my bill? ________
5. Good morning, madam. May I clean your room? ________
6. House keeping. May I help you? ________
7. The traffic outside was too noisy. I couldn't sleep last night. ________
8. Would it be possible for me to check out later? ________

A. The faucet in my bathroom is leaking.
B. I'm sorry to hear that. I'll change another room for you.
C. It depends on the availability of our rooms.
D. Just these four.
E. Yes, but not now. Could you come back in an hour?
F. Sure. The total comes to 680 yuan.
G. Certainly, sir. Where are you going?
H. I'd like to cancel the double room I've reserved.

Ⅲ. Fill in the blanks in the conversation by translating the Chinese into English.

A: Holiday Inn.What can I do for you?

B: Hello! This is Tom. (1) ____________________ (我想给我的老板预订一间双人房).

A: What's the arrival time?

B: On Friday, July 16th.

A: (2) ____________________ (在这个时间段我们确实有可提供的双人间).

The rates for rooms with front view and rear view are different. Which one would you like?

B: What are the rates for a front view room and a rear view room respectively?

A: 160 dollars for the former and 130 dollars for the latter.

B: I'll take a front view room. By the way, (3) ____________________

（这个价位中还包括什么）？

A: The breakfast is complimentary.

B:（4）________________________________（你们酒店有什么娱乐设备吗）？

A: We have gymnasium, billiard room, bridge chess room and dance hall.

B: That sounds great!

A:（5）________________________________（我们期待为您服务）. Bye.

B: Thank you. Bye.

Ⅳ. Remember the useful expressions concerning hotel services .

1. How much do you charge for double and single rooms?
2. Is there a boutique in the hotel?
3. Do you have any vacancies?
4. Could you show it to me, please?
5. Do you need any further information, madam?
6. How do you like our hotel?
7. How long will you be staying?
8. May I help you with your suitcase?

Task 3 Further Development

Passage 1

Pre-reading Task

Read the following statements and tick in the columns under “True”, “False” or “Unsure”.

Statements	True	False	Unsure
There are many reasons for travelling.			
You should be clear about the room types.			
What type of room to choose depends on the number of the people in your group.			
A room with a kitchen/business center/minibar means you need to pay a higher rate.			
There’s no free room service for a luxury suite.			

Types of Rooms in Hotels

No two hotels are the same. What is considered a suite at one hotel could be a basic room at another. Even so, hotel rooms generally fall into distinct categories.

By Beds

While hotels use different phrases to categorize these rooms, they more or less use the same

terminology to indicate how many guests the room can accommodate:

- Single room
- Double room
- Triple room
- Quad room

By Size

Depending on which destinations travelers are visiting, they might find hotel rooms to be smaller than what they've used before. To be sure the size of the room is adequate, request a room that can accommodate the total number of guests:

- Standard room/studio
- Deluxe room
- Joint room
- Suite(mini/junior/master)
- Apartment style

By Amenities

Hotels might also offer different rooms according to the amenities available. While there won't be a name assigned to a room with unique amenities, hotels list which amenities are available depending on the type of room or room size that a traveler requests.

- City view or nature view
- Kitchen
- Handicapped accessible
- Outdoor entrance (cabana or villa)
- Lanai, terrace or balcony
- Minibar
- Hot tub bath
- Pull-out couch
- Business center

By Luxury

Finally, after travelers determine how many beds they want, the size of room they need and the kinds of amenities they want, they can opt for a luxury room. Luxury and size sometimes go hand in hand, but in general, luxury rooms are referred to in a few ways:

- Standard suite
- Executive suite
- Presidential suite

- Penthouse
- Villa

These days, a hotel stay can be anything a traveler wants it to be. It can be family-friendly, private, exotic or simple. No matter what the circumstances surround the trip, travelers deserve to be comfortable when they book their accommodation. Knowing what room types to look for is a good place to start.

New Words

category	['kætəgəri]	*n.*	类别
terminology	[ˌtɜːmɪ'nɒlədʒi]	*n.*	术语
destination	['destɪ'neɪʃn]	*n.*	目的地
adequate	['ædɪkwət]	*adj.*	足够的
amenity	[ə'miːnəti]	*n.*	便利设施
request	[rɪ'kwest]	*v.*	要求
exotic	[ɪg'zɒtɪk]	*adj.*	异国风情的
circumstance	['sɜːkəmstəns]	*n.*	境况

Phrases & Expressions

more or less	或多或少
the number of	……数量
according to	根据；按照
be assigned to	被指定；被分配
go hand in hand	相辅相成；密切相关
be referred to	被提及；涉及
look for	寻找；寻求

Projects

Ⅰ. Fill in the blanks with the proper words given, changing the form if necessary.

accommodation	distinct	indicate	simple
adequate	available	determine	refer

1. This book explains grammar ________ and clearly.
2. Check her ________ before you schedule the meeting.

3. The new law makes no ________ between adults and children.
4. The victims were not ________ to by name.
5. Our hotel can ________ up to 500 guests.
6. The opposition to her plan made her more ________ than ever.
7. His wages are ________ to support three people.
8. A red sky at night often ________ fine weather the next day.

Ⅱ. Translate the following sentences by imitating the examples given, paying attention to the underlined phrases or structures.

1. Depending on which destinations travelers are visiting, they might find hotel rooms to be smaller than what they've used before.
我们是否去野营要看明天的天气。
____.

2. To be sure the size of the room is adequate, request a room that can accommodate the total number of guests.
鱼的数量减少主要归因于过度的捕捞。
____.

3. Hotels might also offer different rooms according to the amenities available.
每个人将根据他的能力获得报酬。
____.

4. No matter what the circumstances surround the trip, travelers deserve to be comfortable when they book their accommodation.
不管你说什么，我将坚持自己的意见。
____.

5. Knowing what room types to look for is a good place to start.
集邮是我的爱好之一。
____.

Passage 2

Hotel Services

Typically, the basic hotel services include guest reception, room service, food service, including restaurants in the hotel, and security. Sometimes in the small hotels the duties of security, a cook and a cleaner are performed by the owner himself.

Other services offered to guests of the hotel can be considered as bonuses. These are the laundry service, massage room, fitness gyms, conference rooms, lock boxes for valuable assets and many other things. These services can be included in the price of the room or paid separately.

Recently, the hotel industry trends towards separating the services sector among hotels. Many hotels nowadays offer recreation for a particular group of tourists. Popular family hotels, hotels for the newlyweds and hotels for people with disabilities – each of them has its unique set of services. For example, in the family hotel clients are offered services of child care and game rooms. In the hotel for the newlyweds there is a special service for weddings. In the hotel for the disabled there is medical support service.

Among the services that are indirectly related to the customers, which means that they are not involved in direct care of guests, are marketing service and bookkeeping of the hotel. Reservation, once considered one of the main hotel services, has become an anachronism today. Nowadays, in order to book a hotel, clients use services of the major tour operators. Online booking through the numerous tourist online services is getting increasingly popular.

New Words

security	[sɪ'kjʊərəti]	*n.*	安全
perform	[pə'fɔ:m]	*v.*	履行
bonus	['bəʊnəs]	*n.*	红利
trend	[trend]	*n.*	趋势
recreation	['rekri'eɪʃn]	*n.*	娱乐
disabled	[dɪs'eɪbld]	*adj.*	失能的；有残疾的
bookkeeping	['bʊkki:pɪŋ]	*n.*	记账
anachronism	[ə'nækrənɪzəm]	*n.*	过时事物

Phrases & Expressions

be considered as	被认为；看作是
be included in	包括在
people with disabilities	残疾人
be involved in	涉及

Check your understanding

Choose the correct answers according to the information given in the passage.

1. Most Hotels these days offer ________ to guests, depending on their specific needs.
   A. specialized services　　B. pay-as-you-go services　　C. random services
2. A common characteristic of some small hotels is that they have ________ number of staff at their disposal.

A. uncountable　　B. unlimited　　C. limited

3. You may need to pay ________ when offered other services in the hotel.

A. separately　　B. extra charges　　C. attention

4. Some of the basic services a hotel should provide for its guests include ________ .

A. clothing and beauty products

B. shelter and feeding

C. dancing and entertainment

5. An easier method of reserving a room these days would be to ________ .

A. reserve online through travel agents

B. travel to the hotel and book a room personally

C. contact other tourists to help you

Task 4 Related Information

Types of Hotels

胶囊旅馆　capsule hotel

汽车旅馆　motel

度假酒店　resort hotel

提供一夜住宿和早餐的旅馆　B & B hotel

青年招待所　youth hostel

豪华酒店　luxury hotel

公寓旅馆　residential hotel

寄宿公寓　boarding house

乡村旅馆　country inn

绿色旅游酒店　green hotel

五星级酒店　five-star hotel

Task 5 Pop Quiz

Ⅰ. Working with words

Match the following words and phrases with their Chinese equivalents.

A. lobby　　B. bellboy

C. baggage check　　D. registration form

E. breakfast voucher F. housekeeper

G. cashier H. headwaiter

I. shuttle bus J. chambermaid

1. 登记表 ( )	6. 酒店大堂 ( )
2. 餐厅领班 ( )	7. 行李托管证 ( )
3. 管家 ( )	8. 行李工 ( )
4. 收银员 ( )	9. 清理房间的女服务员 ( )
5. 专线大巴车 ( )	10. 早餐券 ( )

Ⅱ. Multiple choice

1. Resort hotels distinguish themselves from most other types of hotels by providing ______.
   A. kitchen amenities
   B. food and beverage services
   C. special activities such as horseback riding
2. Which of the following types of hotels are small and usually have the owner living on the premises? ______.
   A. Bed and breakfast hotels B. Conference centers C. Airport hotels
3. Which of the following departments of a hotel has the greatest amount of guest contact? ______.
   A. Housekeeping B. Front desk C. Sales
4. Concierge services are normally part of ______.
   A. the sales department B. the reservations department C. the rooms division
5. When it comes to judging the quality of service, whose expectations are most important? ______.
   A. The owner of the company
   B. The staff member providing the service
   C. The person receiving the service
6. Typical requests handled by the front office include all of the following except ______.
   A. auditing services B. entertainment reservations C. transportation arrangements
7. Nowadays, ______ have pre-arranged bookings of their tour groups in a hotel.
   A. the educational agencies B. the travel agencies C. the airports
8. What type of complaints do guests make when they feel they've been mistreated by the hotel staff? ______.
   A. Service-related complaints B. Unusual complaints C. Attitudinal complaints

Ⅲ. Fill in the blanks with the proper choices given.

Hotel Description

Being close to airport and freeway access makes ___(1)___. The 300-room Embassy Suites offers homey comforts for those on business and vacation alike—suites with separate living and sleeping areas, refrigerators, microwaves, coffeemakers, Wi-Fi access and two TVs. Wake up to a free cooked-to-order breakfast and ___(2)___ and fitness center. A two-hour evening reception offers drinks and appetizers and the ___(3)___ serves American favorites. The property also has a business center ___(4)___. Parking is available for an additional fee. Less than two miles from I-405, the Embassy Suites is ___(5)___. Our guests say the Embassy Suites' location, "brilliant" breakfast and "friendly" staff make it one of their favorite Santa Ana properties.

A. with copy and fax services

B. the on-site restaurant

C. within a half-mile of various restaurants and just three miles from John Wayne Airport

D. enjoy the indoor pool

E. the Embassy Suites a top choice for our guests in the area

UNIT 7 Modern Agricultural Machinery

Task 1 First Sight

1. Match the following pictures with the key words given.

A. speed tiller B. agricultural robot C. turf mower D. agricultural drone

(1)

(2)

(3)

(4)

2. Answer the question by matching the following items.

What roles can agricultural machinery play in modern farming?

□1) make	A. crop yields
□2) relieve	B. labor productivity
□3) liberate	C. worker shortage
□4) raise	D. farming easier
□5) improve	E. manpower

Task 2 Better Acquaintance

Conversation

At the Farm Machinery Show

(Farmer Tom comes to a farm machinery show. He is talking to a salesman to get more information about the machines.)

Tom: Could you explain the advantages of farm machinery?

Salesman: Sure. It's clear that farm machinery can not only make farming easier, but also help improve yields.

Tom: It is reported that there has been a great improvement in this field in recent years.

Salesman: You got the point! Here are some of our farm machines: automatic sprayer, combine harvester, hay rakes, line ripper, rotary hoes and bed formers.

Tom: They look so cool! What's this? I guess it's a fertilizer spreader. I've long been thinking of buying one. Can you tell me more about it?

Salesman: Certainly! It can spread a wide variety of fertilizers and seeds. It's perfect for sporting fields, lawns and pastures.

Tom: What's its spreading width?

Salesman: From 6 to 14 meters.

Tom: And what's its empty machine weight?

Salesman: Here is the manual. Let me have a look..., yes, it's 112 kg.

Tom: Where is it made?

Salesman: It's made in Italy.

Tom: What's the length of its warranty period?

Salesman: We guarantee our products for two years.

Tom: Got it. Thank you so much for your explanation.

Salesman: Here is my business card. Please contact me if you have made your decision.

New Words

machinery	[məˈʃiːnəri]	*n.*	（统称）机器
yield	[jiːld]	*n.*	产量
rotary	[ˈrəʊtəri]	*a.*	旋转的
fertilizer	[ˈfɜːtəlaɪzə(r)]	*n.*	肥料
lawn	[lɔːn]	*n.*	草坪
pasture	[ˈpæstʃə(r)]	*n.*	牧场
width	[wɪdθ]	*n.*	宽度
manual	[ˈmænjuəl]	*n.*	使用手册

Phrases & Expressions

automatic sprayer	自动喷洒机
combine harvester	联合收割机
hay rake	搂草机
line ripper	垄行成型机
rotary hoe	旋转锄地机
bed former	作床机

Projects

Ⅰ. Substitution drills

1. Farm machinery

使农活较为省力.
有助于提高粮食产量.
有助于解放农村劳动力.
可以解决工人短缺问题.
在农场里已经流行起来.

2. Agricultural drones

可以给作物施肥.
可以给作物喷洒农药.
更好地利用水和劳动力.
监控干旱和作物疾病.
可以为农民省钱和保护环境.

3. Small-unmanned aerial vehicles are used in civil fields like

快递运输.
灾难救援.
野生动物观察.
新闻报道.
影视拍摄.

4. Agricultural robots can do

播种.
灌溉.
除草.
施肥.
收割庄稼.

5. This line ripper

可以耕作黏性土壤.
可以耕作岩性土壤.
可以消除土壤板结.
重670千克.
作业宽度为2.8米.

Ⅱ. The following are questions and answers about modern agricultural machinery between two friends. Join the questions 1-8 to the answers A-H. Then act out some of the mini-dialogs with your partner.

1. It's cool to fly a UAV. Is it easy to learn? ________
2. How many types are UAVs divided into in terms of weight? ________
3. What's the length of the warranty period for this hay baler? ________
4. What's the weight of this turf mower? ________
5. What's the working depth of this sub-soil cultivator? ________
6. How many licensed UAV pilots are there in China? ________
7. Are small-unmanned aerial vehicles used in search and rescue operations? ________
8. Can agricultural robots spray weedicides? ________

A. Yes. With their help, the cost of sending a group of people to dangerous places gets reduced.

B. 83 kg.

C. No, it isn't. You have to accept a certain period of theoretical and practical training, pass the examination and then you can get your civilian UAV pilot certificate issued by the CAAC.

D. Of course they can. This can help the user avoid manual spraying which may cause toxicity.

E. We guarantee our products for two years. Within this period, any non-intentional damage will be repaired free of charge.

F. Four, namely micro, light, small and big.

G. It was reported that there had been less than 400 by 2015. But the number must have risen dramatically by now.

H. From 35 to 60 cm.

Ⅲ. Fill in the blanks in the conversation by translating the Chinese into English.

A Real "Mechanized Hired Hand"

（Farmer Tom and his friend Jim are talking about the news report on the prospects of agricultural robots on farms.）

Tom: Hi, Jim! It's reported that in the near future, computer-aided robots will work for us on the farm. They will be able to move and, in some ways,（1）________________________（像人类一样思考）.

Jim: Really? It's exciting! What will they do?

Tom: Scientists are now developing robots that will be able to（2）______________（剪羊毛），（3）______________（驾驶拖拉机），and（4）______________（收获水果）.

Jim: What will they do on my cow farm? Will they milk my cows?

Tom: Of course they will. In addition, when the milking is completed, the robots will automatically check to make sure that（5）______________（牛奶是干净的）. The complete mobilization of the farm is far in the future, but engineers expect that some robots will be used before long.

Jim: So they are hired hands that will never feel tired or complain.

Tom: You are right.

Ⅳ. Remember the useful expressions concerning modern agricultural machinery.

1. I'm very interested in this hay rake. Can you give me more information about it?
2. What's the weight of this farm machine?
3. What's the working depth of it?
4. What's the working width of it?
5. What's the length of the warranty period for it?
6. Can it be used on sticky soil/rocky soil/hilly terrain?
7. Does this machine have an age requirement for its operators?
8. Is it easy/dangerous to operate?

Task 3 Further Development

Passage 1

Pre-reading Task

Read the following statements and tick in the columns under "True", "False" or "Unsure".

Statements	True	False	Unsure
UAVs can be divided into military and civil.			
UAVs are developing rapidly now because they are easy to control and can do plenty of things.			
Farmers use small-unmanned aerial vehicles to view their crop growth from remote distance.			
A licensed pilot should have the ability to plan the flying route of a UAV.			
Small-unmanned aerial vehicles are used in anti-terrorist operations.			

Agricultural Drone

An agricultural drone is an unmanned aerial vehicle applied to farming in order to help increase crop production and monitor crop growth. Through the use of advanced sensors and digital imaging capabilities, farmers are able to use these drones to help them gather a richer picture of their fields. Information gathered from such equipment may prove useful in improving crop yields and farm efficiency.

Agricultural drones let farmers see their fields from the sky. This bird's-eye view can reveal many issues such as irrigation problems, soil variation, pest and fungal infestations. Multispectral images show a near-infrared view as well as a visual spectrum view. The combination shows the farmer the difference between healthy and unhealthy plants, a difference not always clearly visible to the naked eye. Thus, these views can assist in assessing crop growth and production.

Additionally, the drone can survey the crops for the farmer periodically to their liking. Weekly, daily, or even hourly, pictures can show the changes in the crops over time, thus showing possible "trouble spots". Having identified these trouble spots, the farmer can attempt to improve crop management and production.

Types

Remote control type

It must be monitored by the operator at all times through remote control such as radio.

Autonomous expression

It flies autonomously with a navigation device such as a global positioning system.

Fixed wing type

It is suitable for covering a vast area, but a catapult and a runway are necessary when it is not a vertical take-off and landing aircraft.

Rotary wing type

It can take off and land vertically and stop in the air, but its speed is inferior to that of the fixed wing type.

New Words

drone	[drəʊn]	*n.*	无人机
monitor	['mɒnɪtə(r)]	*v.*	监控
sensor	['sensə(r)]	*n.*	传感器
digital	['dɪdʒɪtl]	*adj.*	数码的
irrigation	[ˌɪrɪ'geɪʃn]	*n.*	灌溉
variation	[ˌveəri'eɪʃn]	*n.*	变化
fungal	['fʌŋgl]	*adj.*	真菌的
infestation	[ˌɪnfɛs'teɪʃən]	*n.*	横行；侵扰
multispectral	[ˌmʌlti'spektrəl]	*adj.*	多光谱
visual	['vɪʒuəl]	*adj.*	视觉的
spectrum	['spektrəm]	*n.*	光谱
assess	[ə'ses]	*v.*	评估
periodically	[ˌpɪərɪ'ɒdɪkəli]	*adv.*	定期地
identify	[aɪ'dentɪfaɪ]	*v.*	发现
attempt	[ə'tempt]	*v.*	努力
autonomous	[ɔː'tɒnəməs]	*adj.*	自主的
navigation	[ˌnævɪ'geɪʃn]	*n.*	导航
catapult	['kætəpʌlt]	*n.*	弹射器
take-off	['teɪk ɒf]	*n.*	（飞机）起飞
inferior	[ɪn'fɪəriə(r)]	*adj.*	比不上的

Phrases & Expressions

attempt to do	尝试做某事
take off	（飞机）起飞
be inferior to	逊色于
agricultural drone	农业无人机
UAV（Unmanned Aerial Vehicle）	无人机
GPS（Global Positioning System）	全球定位系统

Projects

Ⅰ. Fill in the blanks with the proper words given, changing the form if necessary.

efficiency	irrigation	variation	infestation
assess	identify	inferior	navigation

1. If children were made to feel ________ to other children, their confidence declined.
2. Parts of this country are suffering from an ________ of oriental fruit flies.
3. There are many ways to increase agricultural ________ in the poorer areas of the world.
4. Do these two ________ systems play different roles?
5. A ________ is a change or slight difference in a level, amount, or quantity.
6. ________ is needed to make crops grow in dry regions.
7. We are trying to ________ how well the system works.
8. Scientists have ________ a link between diet and cancer.

Ⅱ. Translate the following sentences by imitating the examples given, paying attention to the underlined phrases and structures.

1. Information gathered from such equipment may prove useful in improving crop yields and farm efficiency.

   停在门口的汽车是王教授的。

   ____.

2. This bird's-eye view can reveal many issues such as irrigation problems, soil variation, pest and fungal infestations.

   应该鼓励儿童去培养自己在课外的爱好，比如音乐和体育。

   ____.

3. Multispectral images show a near-infrared view as well as a visual spectrum view.

   他不仅在乡下有一栋房子，在城里也有一栋。

   ____.

4. Weekly, daily, or even hourly, pictures can show the changes in the crops over time, thus showing possible "trouble spots".
   农业迅速发展，从而为轻工业提供了充足的原料。
   ____.

5. Having identified these trouble spots, the farmer can attempt to improve crop management and production.
   他浇完花后，开始割草。
   ____.

Passage 2

Agricultural Robots Mitigate Worker Shortages

Across the world, farmers are aging. The shortage of people willing to work on the farm is growing chronic everywhere. In Japan alone, the number of farmers dropped from 2.2 million in 2004 to 1.7 million in 2014.

In addition, most young people don't consider farming an attractive profession, and immigration policies around the world are making it difficult to obtain migrant workers.

From UAVs to driverless vehicles, a new breed of robotic workers is coming to the farm.

Thanks to improvements in perception and manipulation, agriculture robots can pick grapes, lettuce, or strawberries faster and with as much delicacy as humans.

As migrant workers from Eastern Europe leave or avoid the U.K. in the wake of the so-called Brexit from the EU, robots will pick strawberries.

In the U.S., concerns about President Donald Trump's plans for a border wall are affecting the seasonal farm workers from Mexico. So one Michigan farm has turned to agricultural robots that can pick apples and that are three times more productive than humans.

In Japan, Spread Co. plans to use robots to produce lettuce. Its vertical farm will be fully automated. The only job for humans is to plant seeds. Robots will do everything else, from watering and replanting to harvesting.

Without robots, Spread's farm can produce 21,000 heads of lettuce a day. With robots, the volume will go up to 51,000 lettuce heads a day.

New Words

mitigate	[ˈmɪtɪɡeɪt]	*v.*	减轻
shortage	[ˈʃɔːtɪdʒ]	*n.*	不足
chronic	[ˈkrɒnɪk]	*adj.*	长期的

profession	[prəˈfeʃn]	*n.*	职业
breed	[bri:d]	*n.*	类型
perception	[pəˈsepʃn]	*n.*	认知
manipulation	[məˌnɪpjʊˈleɪʃ(ə)n]	*n.*	操作
lettuce	[ˈletɪs]	*n.*	莴苣
delicacy	[ˈdelɪkəsi]	*n.*	审慎；周到
productive	[prəˈdʌktɪv]	*adj.*	多产的

Phrases & Expressions

in the wake of	随着……而来
immigration policy	移民政策
migrant workers	外来务工者
Brexit	英国脱欧
vertical farm	垂直农场

Proper Nouns

Michigan	[ˈmɪʃəgən]	*n.*	密歇根州

Check your understanding

Choose the correct answers according to the information given in the passage.

1. The number of people willing to work on farms ________ across the world.
   A. is rising　　B. is dropping　　C. remains the same
2. Young people tend to have ________ attitudes towards working on farms.
   A. positive　　B. negative　　C. neutral
3. According to the passage, migrant workers picking fruit on farms in the UK and the U.S. are mainly from ________.
   A. Eastern Europe and Mexico
   B. Africa and Asia
   C. Southern Europe and Mexico
4. ________ will take the place of migrant workers to pick fruit on the farm soon.
   A. Part-timers　　B. The farmers themselves　　C. Agricultural robots
5. Robots can do the following EXCEPT ________ on the vertical farm of Spread Co. in Japan.
   A. seeding　　B. harvesting　　C. watering

Task 4　Related Information

Understand the product information about the bed former.

Maximum horse power: 80 HP（59 kW）

4 speed gearbox for 540 rpm PTO（176, 197, 221, 248 rpm）

3 point hitch cat. 2

Structure frame completely welded（high-strength thick steel）

C-shaped blades 80 mm×8 mm（6 blades each flange leaning inwards）with bolted double flanges （rotor+blades d.550）

Rear tailgate with spring cushioned comb

Working depth 25 cm

Standard rear bed former

Gears side drive

Drive shaft with cam clutch

Task 5　Pop Quiz

Ⅰ. Working with words

Match the following English technical terms with their Chinese equivalents.

A. licensed UAV pilot　　B. combine harvester

C. hay rake　　D. fertilizer spreader

E. working width　　F. working depth

G. civilian use　　H. military use

I. turf mower　　J. agricultural drone

K. agricultural robot　　L. pesticide spraying

1. 作业深度	(　　)	6. 军用	(　　)
2. 农业无人机	(　　)	7. 农药喷洒	(　　)
3. 执业无人机飞手	(　　)	8. 施肥机	(　　)
4. 作业宽度	(　　)	9. 民用	(　　)
5. 联合收割机	(　　)	10. 农业机器人	(　　)

Ⅱ. Multiple choice

1. Tobacco is a crop that is traditionally harvested ________.

   A. by machinery　　B. by hand　　C. both A and B

2. Commercial dairy farms usually use _______ in order to produce clean, fresh milk for consumers.

   A. machinery　　B. manual labor　　C. both A and B

3. Some ______ are almost completely automated, while ______ may pride themselves with doing much of the work by hand, and thus providing a very high-quality product.

   A. large farms; small farms　　B. small farms; large farms　　C. Unsure

4. About 80 to 90 percent of drone crashes are caused by _______.

   A. altitude　　B. speed　　C. operators' mistakes

5. To become a UAV pilot, you must get a civilian UAV pilot certificate issued by the _______.

   A. ACCA　　B. CAAC　　C. CACA

6. Small-unmanned aerial vehicle is a small aircraft, which flies without direct human contact and is controlled by a _______.

   A. steering wheel　　B. battery　　C. remote control

7. While the old-time farms depended on ______ power, and modern farms depend on ______ power, farms of the future will depend on ______ power.

   A. machine; horse; computer　　B. horse; machine; computer　　C. computer, machine; horse

8. Agricultural robots on farms can __________.

   A. relieve worker shortages　　B. improve yields　　C. both A and B

Ⅲ. Fill in the blanks with the proper choices given.

Farm workers who learn from the start how to run all of the pieces of equipment on a farm increase their chances of being offered full-time hours and pay raises. However, ___(1)___:

Often, there is an age minimum. While some farms hire high school students, especially for seasonal farm work, most machinery requires that you are at least 18 years old to run it. In some cases, ___(2)___.

Equipment is expensive. Unless you have training, you could break equipment, ___(3)___. On-the-job training is extremely important, and don't be afraid to ask questions.

If you use machinery improperly, you could also hurt the animals. Most machinery is built with safety triggers to avoid hurting animals and workers alike, ___(4)___. Once again, make sure you have been trained before you attempt running a piece of machinery.

Even with training, machinery is dangerous. Every year, farm workers are killed when machinery rolls over or catches on fire. Many pieces of farm equipment also have automated parts with sharp edges or forceful movements, and ___(5)___.

A. but this isn't always the case

B. costing the farmer thousands of dollars and costing you your job

C. it is easy for farmers to lose digits or limbs

D. there are a few things to consider when it comes to running machinery

E. that age requirement may be even higher

UNIT 8 Modern Crop Farming

Task 1 First Sight

1. Match the following pictures with the key words given.

A. drip irrigation B. insect-proof glue board

C. soilless cultivation D. aquaponics

(1)

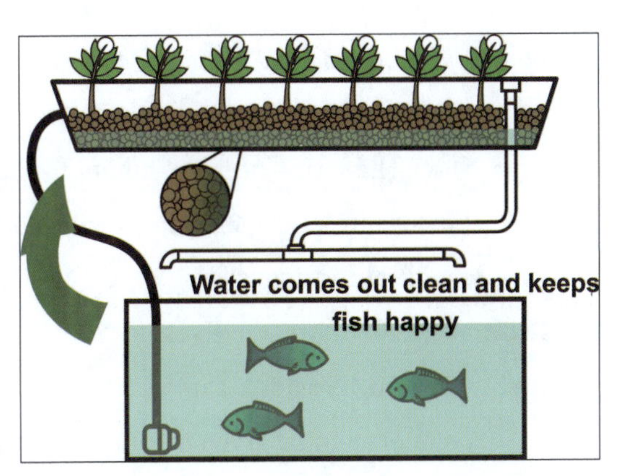

(2)

(3)

(4)

2. Answer the question by matching the following items.

What are the key factors for sustainable farming?

□1）crop	A. crops
□2）vertical	B. robots
□3）organic	C. rotation
□4）agricultural	D. greenhouses
□5）AI	E. farms

Task 2 Better Acquaintance

Conversation

Are Organic Foods Really Healthier?

（Tom is consulting his friend Jim about organic foods.）

Tom: I've noticed that some produce labeled "organic" at supermarkets is several times more expensive than the non-organic. Could you explain what "organic" means, Jim?

Jim: Sure. The term "organic" refers to the way agricultural products are grown and processed. For example, organic crops must be grown without the use of synthetic pesticides and chemical fertilizers.

Tom: What about organic livestock?

Jim: Organic livestock raised for meat, eggs, and dairy products must have access to the outdoors and be given organic feed. They are not given antibiotics, growth hormones or any animal by-products.

Tom: What are the benefits of organic vegetables and fruits?

Jim: They often have some beneficial nutrients like antioxidants. What's more, they are much fresher because they are often produced on smaller farms near where they are sold.

Tom: So organic meat and milk are more nutritious as well. Is that right?

Jim: Absolutely. Results of a study show that levels of certain nutrients, including omega-3 fatty acids, are up to 50 percent higher in organic meat and milk than in conventionally raised versions.

Tom: I see. Organic foods are much healthier and worth the expense. For the sake of our health, organic foods should be our first choice.

New Words

label	[ˈleɪbl]	*v.*	用标签标明
organic	[ɔːˈgænɪk]	*adj.*	有机的
synthetic	[sɪnˈθetɪk]	*adj.*	合成的

pesticide	[ˈpestɪsaɪd]	*n.*	杀虫剂
livestock	[ˈlaɪvstɒk]	*n.*	家畜
hormone	[ˈhɔːməʊn]	*n.*	激素
by-product	[ˈbaɪˌprɒdʌkt]	*n.*	副产品
benefit	[ˈbenɪfɪt]	*n.*	优势
nutrient	[ˈnjuːtriənt]	*n.*	营养素
antioxidant	[ˌæntiˈɒksɪdənt]	*n.*	抗氧化物质
conventionally	[kənˈvenʃənəli]	*adv.*	传统地

Phrases & Expressions

have access to	有机会或权利使用
for the sake of	出于……考虑
omega-3 fatty acid	ω-3不饱和脂肪酸

Projects

Ⅰ. Substitution drills

1. Organic foods

不含抗生素和生长激素.
几乎不含杀虫剂.
无转基因成分.
更新鲜.
含有更多有益的营养成分.

2.

黄瓜
芹菜
葡萄
菠菜
草莓

has the highest pesticide levels, so it is BEST to buy organic.

3. Fruits and vegetables you DON'T need to buy organic are

蘑菇.
洋葱.
甜玉米.
芒果.
甘薯.

4. Soilless cultivation

把沙漠变成菜园. 特色是更优质、更高产. 适合种植蔬菜、水果和草本植物. 并不仅仅是另一种生产农产品的方式. 可缓解由世界人口增长导致的土地短缺.

5.

内部 外部 窗玻璃 铝质框架 木质框架

of the greenhouse should be regularly cleaned and maintained.

Ⅱ. The following are questions and answers about modern crop farming between two friends. Join the questions 1-8 to the answers A-H. Then act out some of the mini-dialogs with your partner.

1. How are weeds controlled on organic farms and non-organic ones? ________
2. How are diseases prevented on organic livestock farms and non-organic ones? ________
3. Does organic mean pesticide-free? ________
4. What are the possible risks of pesticides? ________
5. Washing and peeling produce can get rid of pesticides. Is that true? ________
6. What are the majority of greenhouse structures made of? ________
7. There are more plastic greenhouses than glass greenhouses in Japan. Why? ________
8. What is the main attraction of soilless cultivation? ________

A. No, it doesn't. Despite popular belief, organic farms do use pesticides. The difference is that they only use naturally-derived pesticides, which are believed to be less toxic.

B. Because earthquakes occur frequently in Japan, which is a factor that limits the development of glass greenhouses.

C. They are made of plastic followed by glass and other materials.

D. Certainly not. Rinsing reduces but does not eliminate pesticides. Peeling sometimes helps, but valuable nutrients often go down the drain with the skin. The best approach: eat a varied diet, wash and scrub all produce thoroughly, and buy organic when possible.

E. They are prevented with natural methods such as clean housing, rotational grazing and healthy diet on organic livestock farms while on non-organic ones, antibiotics and medications are used.

F. Higher yields and higher quality, I think.

G. Some studies have indicated that the use of pesticides even at low doses can increase the risk of certain cancers, such as leukemia, brain tumor and breast cancer.

H. They are controlled naturally (crop rotation, hand weeding, mulching, and tilling) on organic farms while on non-organic ones, they are controlled with chemical herbicides.

Ⅲ. Fill in the blanks in the conversation by translating the Chinese into English.

Greenhouse Drip Irrigation

（Tom comes to visit his friend Jim's greenhouse and wants to know more about the drip irrigation system.）

Tom: The drip irrigation seems to work well in your greenhouse, but I do have some questions about it. Would you mind explaining them to me?

Jim: Of course not.（1）____________________（请继续）.

Tom: What is drip irrigation?

Jim: Drip irrigation, also known as micro irrigation or localized irrigation, is an irrigation method that saves water and fertilizer by（2）__________________________________（把水缓慢滴入植物的根部）.

Tom: How does it work?

Jim: It's controlled by a battery-powered water timer, which（3）________________________（与水龙头相连） and allows you to choose the times and duration of irrigation. The water timer is called（4）________________________（滴灌系统的大脑）.

Tom: What are its other benefits besides highly efficient use of water and fertilizer?

Jim: It's easy to install and design and the cost is relatively low. What's more, drip irrigation can reduce diseases due to（5）__________________________（某些植物湿度过大）.

Tom: Got it. By the way, what about its labor cost?

Jim: It's much lower than that of any other irrigation method.

Ⅳ. Remember the useful expressions concerning modern crop farming.

1. Are organic foods really healthier?
2. Organic farming is better for the environment.
3. Does organic mean pesticide-free?
4. What are the possible risks of pesticides?
5. Are GMOs safe?
6. Organic foods often have more beneficial nutrients, such as antioxidants.
7. Leafy greens have been growing in popularity in recent years because of the nutritional benefits and convenience as a ready-to-eat (RTE) product.
8. Soilless cultivation optimizes yields both of the plants and the growers who can produce higher quality and quantity, regardless of the seasons.

Task 3 Further Development

Passage 1

Pre-reading Task

Read the following questions and tick in the columns under "Yes", "No" or "Unsure".

Questions	Yes	No	Unsure
Have you ever seen or heard about urban vertical farms?			
Are foods from vertical farms organic without pesticides or chemical fertilizers?			
Can vertical farms create a sustainable environment for city centers?			
Can vertical farms guarantee the yields regardless of the drought, flood and plant diseases?			
Can vertical farms save farmland and help keep the balance of ecosystem?			

What Is a Vertical Farm?

A vertical farm is a large tower with multiple levels, each with soil and crops. Because a vertical farm is a tightly controlled environment, high yields can be attained. With crops growing indoors, continuous farming will occur under a diversity of climatological or ecological circumstances.

A vertical farm can be an independently functioning ecosystem, as it is separated from the outside. Bugs will have to be kept out indefinitely, and plants will need adequate ventilation. Any rotting organic material needs to be recycled or cheaply disposed of. The more efficient the structure is, the less maintenance is required, and ultimately the greater return on investment for its owners. In the more distant future, the vertical farm will be entirely automated.

Indoor farming has been used before, primarily for low-sized crops like herbs and tomatoes, but a vertical farm project will scale that up and allow the production of bigger crops such as wheat. One limitation will be the inability to raise livestock, unless it is highly customized for the indoor environment. In the long term, there is a strong incentive to move away from livestock due to the large amount of food required to feed an animal until it reaches a point at which it can be slaughtered profitably. With success, vertical farm techniques can also be applied to farming in space or on other planets. This idea is often heard in connection with the phrase "urban sustainability".

New Words

vertical	[ˈvɜːtɪkl]	*adj.*	垂直的
multiple	[ˈmʌltɪpl]	*adj.*	多种多样的
attain	[əˈteɪn]	*v.*	达到
diversity	[daɪˈvɜːsəti]	*n.*	多样性
climatological	[ˌklaɪmətəˈlɒdʒɪkl]	*adj.*	与气候学有关的
indefinitely	[ɪnˈdefɪnətli]	*adv.*	无限期地
ventilation	[ˌventɪˈleɪʃən]	*n.*	通风
rot	[rɒt]	*v.*	腐烂
maintenance	[ˈmeɪntənəns]	*n.*	维护
primarily	[praɪˈmerəli]	*adv.*	主要地
customize	[ˈkʌstəmaɪz]	*v.*	定制
incentive	[ɪnˈsentɪv]	*n.*	激励
slaughter	[ˈslɔːtə]	*v.*	屠宰
sustainability	[səsˌteɪnəˈbɪlɪti]	*n.*	可持续性

Phrases & Expressions

scale...up	扩大
in the long term	从长远来看
move away from	远离
due to	由于
apply...to	应用于……
in connection with	与……有关

Projects

Ⅰ. Fill in the blanks with the proper words given, changing the form if necessary.

attain	diversity	indefinitely	ventilation
maintenance	primarily	customize	sustainability

1. If you ________ a room or building, you allow fresh air to get into it.
2. The trial was postponed ________.
3. The school pays for heating and the ________ of the buildings.
4. There is a need for greater ________ and choice in education.
5. If you ________ something, you gain it or achieve it, often after a lot of effort.

6. But this country still has concerns about exports and the ________ of the global recovery.

7. The body is made up ________ of bone, muscle and fat.

8. If you ________ something, you change its appearance or features to suit your tastes or needs.

Ⅱ. Translate the following sentences by using the phrases given.

1. 他们无法处理掉自己产生的有害废料。（dispose of）
2. 我们考虑在明年扩大培训计划。（scale up）
3. 你也许换工作，或者被迫离开熟悉的领域。（move away from）
4. 这项技术被应用于救援行动。（apply...to）
5. 我的问题与我们昨天的讨论有关。（in connection with）

Passage 2

Soilless Cultivation Is the Future

Soilless cultivation is much more than just another way of growing produce. It is a misconception to believe that the issue of workable land is the main reason growers are drawn to soilless cultivation.

Higher yields and higher quality is the main attraction. This higher level of cultivation technology is available and successful for flowers, vegetables and herb growers alike.

The name of the game in soilless culture is Control:

—Control over the pH & EC in the root zone

—Control over the water & nutrients uptake

—Control over the soil quality

—Control over off-season produce cultivation

This precision leads to optimal growing conditions and betters the yields in all fields, during all seasons.

The leading substrate in the world today is coco peat. It is organic, has a good air/water ratio and is readily available at reasonable prices.

Soilless cultivation optimizes yields both of the plants and the growers who can produce higher quality and quantity, regardless of the seasons, while recycling and reusing the water and nutrients as well as the system itself.

The growers' understanding of the plant and its needs grows and is enhanced by the knowledge acquired with the soilless cultivation technology. Soilless cultivation is the future for an ever growing world population who can not afford to continuously plant and replant in the soil or risk their yields to soil deterioration.

New Words

misconception [ˌmɪskən'sepʃn] *n.* 错误认识

attraction	[ə'trækʃn]	*n.*	有吸引力的事
herb	[hɜ:b]	*n.*	草本植物
uptake	['ʌpteɪk]	*n.*	吸收
precision	[prɪ'sɪʒn]	*n.*	准确
optimal	['ɒptɪməl]	*adj.*	最佳的
substrate	['sʌbstreɪt]	*n.*	基底
ratio	['reɪʃiəʊ]	*n.*	比例
optimize	['ɒptɪmaɪz]	*v.*	使最优化
quantity	['kwɒntəti]	*n.*	数量
acquire	[ə'kwaɪə]	*v.*	获得
deterioration	[dɪˌtɪərɪə'reɪʃən]	*n.*	退化

Phrases & Expressions

more than	不仅仅是
regardless of	不论
coco peat	椰纤土

Check your understanding

Decide whether the following statements are true or false according to the information given in the passage.

1. Soilless cultivation is widely popularized mainly because it addresses the problem of workable land shortage.
2. The greatest advantage of soilless cultivation is higher yields and higher quality.
3. "Control" is the key word in soilless cultivation.
4. Coco peat is the leading substrate for soilless cultivation and is high in price.
5. Soilless cultivation guarantees all-year-round production.

Task 4 Related Information

Size and Technical Specifications of Polytunnel Greenhouse-18 m^2

Product Name	Polytunnel greenhouse-18 m^2
Dimensions	600 cm(L)*300 cm(W)*200 cm(H)
Frame Material	Steel tube; Galvanized for surface; Diameter: 25 mm Thickness: 0.7 mm
Cover Material	Green PE mesh fabric 140 g/m^2
Connector	Steel screw
Packaging	Brown box + Color label
Carton Size	122.5 cm×40 cm×30 cm
N.W./G.W.	40/43 kg
PC S/C TN	1 PC

Task 5 Pop Quiz

Ⅰ. Working with words

Match the following English technical terms with their Chinese equivalents.

A. heated greenhouse
B. unheated greenhouse
C. heating system
D. ventilation system
E. compost
F. drip irrigation
G. in-season vegetables
H. off-season vegetables
I. synthetic pesticides
J. aquaponics cycle
K. organic dairy products
L. growth hormone

1. 反季节蔬菜 (　　)	6. 合成农药 (　　)
2. 时令蔬菜 (　　)	7. 有机乳制品 (　　)
3. 通风系统 (　　)	8. 加温温室 (　　)
4. 生长激素 (　　)	9. 滴灌 (　　)
5. 鱼菜共生系统 (　　)	10. 堆肥 (　　)

Ⅱ. Multiple choice

1. ________ greenhouse is the mainstream greenhouse type in Europe.

   A. Plastic　　B. Glass　　C. both A and B

2. _______ is a factor that limits the development of glass greenhouse in Japan.

   A. Tsunami　　B. Landslide　　C. Earthquake

3. In the Netherlands greenhouse is mainly used for planting________ while in the rest of the world, the greenhouse is mainly used for planting _______.

   A. flowers; vegetables and fruits

   B. vegetables and fruits; flowers

   C. flowers; ground cover plants

4. Compared to aluminium greenhouses, wooden framed ones generally need ________ maintenance.

   A. more　　B. less　　C. the same

5. Compared to conventional crop production, greenhouse horticulture production provides ________ food.

   A. cheaper and fresher　　B. safer and healthier　　C. more expensive and fresher

6. Organic crops are grown with the following EXCEPT ________ .

   A. manure　　B. chemical fertilizer　　C. compost

7. Organic foods often have ________ beneficial nutrients and ________ pesticides.

   A. more; fewer　　B. fewer; more　　C. fewer; fewer

8. The adoption of micro-spray irrigation technology in the greenhouses helps local farmers ________ .

   A. save money　　B. save water　　C. save electricity

Ⅲ. Fill in the blanks with the proper choices given.

Aquaponics

Aquaponics is a soil-free method to practice ___(1)___ that has gained popularity in the US. This technology involves a vegetable production system that integrates soil-free cultivation and aquaculture. Aquaponics is a marriage of ___(2)___ . Fish are raised in tanks via a filtration system, which is linked to ___(3)___ where fruits and vegetables flourish. The fish waste and uneaten foods are broken down into ammonia and nitrates by beneficial bacteria. Nitrates serve as fertilizer to the plants and flourish them as a result. The plants serve as filters and hence purify the water which ___(4)___. This process renders the aquaponics system ___(5)___ since spent water and nutrients are reused.

A. water growing beds　　B. environmental-friendly

C. sustainable agriculture　　D. fish production and soilless vegetable production

E. circulates into the fish tanks

UNIT 9 Modern Animal Farming

Task 1 First Sight

1. Match the following pictures with the key words given.

A. disease diagnosis   B. domestic pigs

C. veterinary drugs   D. free-range chickens

(1)

(2)

(3)

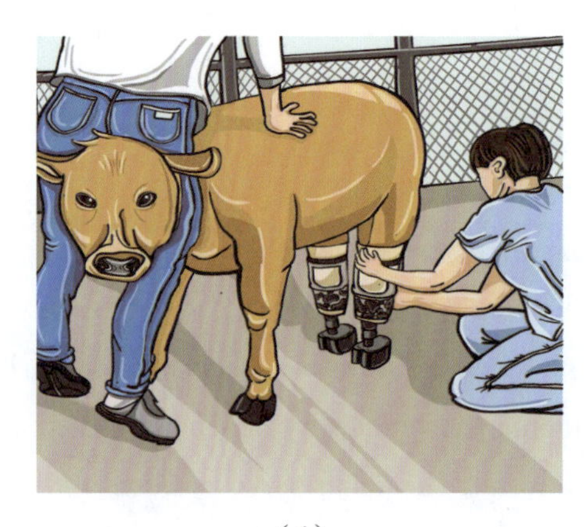

(4)

2. Answer the question by matching the following items.

What are the causes of poultry diseases?

□1）infectious	A. cages
□2）non-nutritional	B. viruses
□3）packed	C. immunity
□4）low	D. injury
□5）physical	E. feeding

Task 2 Better Acquaintance

Conversation

What Are "Free-Range Chickens"?

（Judy is consulting James about free-range chickens.）

Judy: Hey, James. What a large farm! Do you keep any chickens here?

James: Yes, there are lots of chickens including laying hens and broiler chickens. We don't have any battery chickens.

Judy: Why not?

James: Because battery chickens are raised in confinement buildings, with feed and water piped in. Compared with free-range chickens, these battery chickens never see sunshine.

Judy: Can you tell me what free-range chickens are?

James: Free-range chickens are chickens that are allowed frequent access to the outdoors, with plenty of fresh vegetables, sunshine and room to exercise. Moreover, they have not been given any chemicals, so free-range chickens may taste better and make healthier eggs.

Judy: What are the disadvantages of free-range chickens?

James: In fact, free-range chickens are easy to get sick. Besides, it is difficult to manage them.

Judy: I'm doing a school project on the comparison between free-range chickens and battery ones. Would you please tell me more information about free-range chickens?

James: Well, let's go to my office. I'll tell you the details.

New Words

range	[reɪndʒ]	*n.*	（变动或浮动的）范围
hen	[hen]	*n.*	母鸡
broiler	[ˈbrɔɪlə(r)]	*n.*	肉鸡

battery	[ˈbætri]	*n.*	电池
moreover	[mɔːrˈəʊvə(r)]	*adv.*	此外
chemical	[ˈkemɪkl]	*n.*	化学品
disadvantage	[ˌdɪsədˈvɑːntɪdʒ]	*n.*	缺点

Phrases & Expressions

pipe...in	用管道传送……
compared with	与……相比
access to	使用……的机会（权利）
free-range chicken	散养鸡
battery chicken	电池鸡（层架式鸡笼所养的鸡）
laying hen	蛋鸡
broiler chicken	肉鸡

Projects

Ⅰ. Substitution drills

1. Development of aquaculture should focus on

高效.
优质.
生态.
健康.
安全.

2. Suitable housing is needed for keeping the cattle away from

暴雪.
雨水.
大风.
高温.
严寒.

3. Common store-bought chicken is

新鲜的.
自然的.
散养的.
无激素的.
无抗生素的.

4. Poultry farmers should pay attention to

鸡舍管理.
供水设施.
照明系统.
喂养方式.
疾病预防.

5. Some popular poultry breeds for farming include

鸡.
鸭.
鹅.
火鸡.
鸽子.

Ⅱ. The following are questions and answers about modern animal farming between two people. Join the questions 1-8 to the answers A-H. Then act out some of the mini-dialogs with your partner.

1. I really need more knowledge about pig farming, for I am interested in the business. Can you tell me something about it? ________
2. I want to start poultry farming business. Can you give me some suggestions? ________
3. What do you think of veterinary drug residues? ________
4. What kind of services does Aquaculture Farming Technology（AFT）provide? ________
5. Which is the proper method of raising chickens, free range or cage-raised? ________
6. Why is African swine fever（ASF）a worldwide concern? ________
7. What do farmers need to do to improve milk production in dairy farming? ________
8. Do you have any equipment used in chicken houses? ________

A. As far as I know, veterinary drugs are dangerous to human health if their residues are allowed to enter the food chain.

B. They need to learn about dairy or cattle management, proper breeding, balanced diet, disease prevention, primary care and modern farming methods to get better production.

C. Pig farming is a good industry requiring planning and good management. You can contact me by e-mail for details.

D. No problem. If you want to start the chicken raising business, first, you should learn from the nearest chicken farmer.

E. ASF is a worldwide concern as it can spread quickly and cause significant economic losses.

F. It offers a wide range of services in the field of shrimp aquaculture.

G. Yes, we do. Our equipment is popular all over the world. If you need any help, please don't hesitate to contact me.

H. Raising your local breeds in free range system will be the most convenient.

Ⅲ. Fill in the blanks in the conversation by translating the Chinese into English.

Humane Slaughter Method

(Two friends are talking about a new slaughter method.)

Mr. Chen: I've heard the news that McDonald's is going to use a new chicken slaughter method.

Mike: Really? What is it?

Mr. Chen: Controlled-atmosphere killing, or CAK, according to the animal rights group PETA.

Mike: (1) ______________________ (这是什么意思)?

Mr. Chen: It means putting the chickens to sleep quickly and painlessly by removing oxygen from crates that carry chickens and (2) ______________________ (用其他气体替换).

Mike: Sounds good. I think that chickens should be slaughtered (3) ______________ (以人道的方式).

Mr. Chen: Major slaughterhouses have adopted this method. For example, they stun chickens by dipping their heads into electrified water. Many experts are in favor of these methods and other humane practices. They also offer similar suggestions for (4) ______________________ (猪、牛、羊的饲养、运输和屠宰).

Mike: (5) ______________________ (我完全同意). Thanks for your sharing the information.

Mr. Chen: My pleasure.

Ⅳ. Remember the useful expressions concerning modern animal farming.

1. Nowadays, poultry raising has become a profitable and potential industry.
2. Poultry birds can be raised in both free range and indoor production systems.
3. What does "Free Range" mean?
4. According to the effectiveness and production of milk and meat, the cattle breeds are of four types.
5. What are the symptoms of foot-and-mouth disease (FMD)?
6. Animal health or veterinary drug products are the materials intended for animals rather than humans, including drugs used in animal feeds.
7. Nutritional additives include minerals, vitamins and amino acids, which are mainly used to meet the nutrients needed for the growth and development of livestock and poultry.
8. We can make poultry farming more profitable by applying modern science and technology to it.

Task 3 Further Development

Passage 1

Pre-reading Task

Read the following statements and tick in the columns under "True", "False" or "Unsure".

Statements	True	False	Unsure
Artificial intelligence can solve the problem of antibiotic resistance in both farm animals and humans.			
Artificial intelligence is reducing the need for antibiotic treatment in chickens now.			
Antibiotic resistance is a worldwide problem.			
Large numbers of people and animals are given antibiotics when they don't need them.			
The plans to use artificial intelligence have been successful.			

Fighting Drug Resistance via AI

Scientists have announced plans to use artificial intelligence on chicken farms in order to combat the problem of antibiotic resistance in both farm animals and humans.

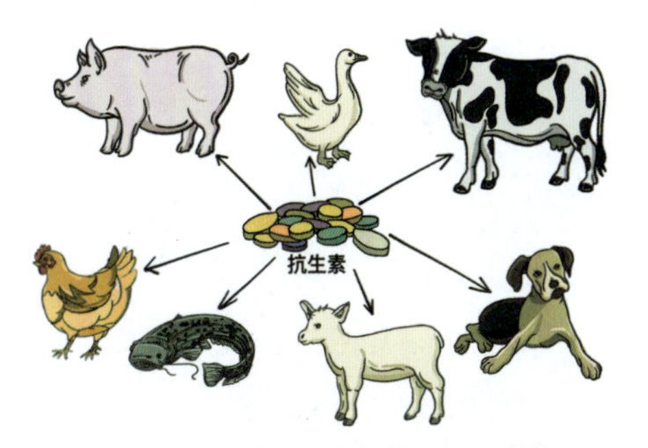

The new initiative will use machine learning to find ways to track and prevent diseases on poultry farms, reducing the need for antibiotic treatment in chickens and therefore lowering the risk of antibiotic-resistant bacteria transferring to people.

Antibiotic resistance is a worldwide problem and it's getting worse and worse. Some of these superbugs are resistant to everything, and we don't know how to treat them. On farms, superbugs are not confined to animals, and they can spread to humans and to the environment. Therefore, it's an exponential spread. If we don't understand how to stop this, it's going to be really bad.

Antibiotics work by disrupting function in certain parts of a bacterial cell. Bacteria become resistant to antibiotics through genetic mutations that alter those areas of the cell, meaning the medication can no longer target them.

The more a strain of bacteria is exposed to an antibiotic, the more likely it is to become resistant. Large numbers of people and animals are given antibiotics when they don't need them, so reducing unnecessary consumption is crucial in the fight against so-called superbugs.

When you have a large-scale data set, the human mind can't cope with that, for it's too complex. We need something that is able to understand the relationship across a big amount of information. If successful, these methods should be transferable to other farm studies in the world.

New Words

announce	[ə'naʊns]	*v.*	宣布
combat	['kɒmbæt]	*v.*	战斗
initiative	[ɪ'nɪʃətɪv]	*n.*	倡议
track	[træk]	*v.*	跟踪
poultry	['pəʊltri]	*n.*	家禽
lower	['ləʊə(r)]	*v.*	减少
transfer	[træns'fɜ:(r)]	*v.*	转移
superbug	['su:pəbʌg]	*n.*	超级细菌（抗生素不能轻易杀死）
exponential	[ˌekspə'nenʃl]	*adj.*	越来越快的
disrupt	[dɪs'rʌpt]	*v.*	打乱
function	['fʌŋkʃn]	*n.*	功能
mutation	[mju:'teɪʃn]	*n.*	（生物物种的）变异
medication	[ˌmedɪ'keɪʃn]	*n.*	药物
target	['tɑ:gɪt]	*v.*	把……作为攻击目标
crucial	['kru:ʃl]	*adj.*	关键性的
strain	[streɪn]	*n.*	（植物、动物的）品种
consumption	[kən'sʌmpʃn]	*n.*	（能量、食物或材料的）消耗

Phrases & Expressions

artificial intelligence（AI）	人工智能
antibiotic resistance	抗生素抗药性
be confined to	局限于……
genetic mutation	基因突变
a strain of	一种
so-called	所谓的
large-scale	大规模的
data set	数据集

Projects

Ⅰ. Fill in the blanks with the proper words and expressions given, changing the form if necessary.

intelligence	treat	announce	alter
resistant	consumption	transfer	confine

1. Negotiations in the past two months have centered on ________ the balance of the offer, as well as raising the price.
2. How can I ________ money from my bank account to his?
3. His genius was not ________ to the decoration of buildings.
4. The average daily ________ of fruit and vegetables is around 200 grams.
5. Most people of average ________ would find this task quite difficult.
6. Some people are very ________ to the idea of exercise.
7. An official ________ of their plans is expected to follow early in the New Year.
8. Many patients are not getting the medical ________ they need.

Ⅱ. Translate the following sentences by imitating the examples given, paying attention to the underlined phrases or structures.

1. Scientists have announced plans to use artificial intelligence on chicken farms in order to combat the problem of antibiotic resistance in both farm animals and humans.
   我们学习一种语言是为了交流思想。
   ____.
2. Antibiotic resistance is a worldwide problem and it's getting worse and worse.
   这里优美的风景吸引了越来越多的旅游者。
   ____.
3. Antibiotics work by disrupting function in certain parts of a bacterial cell.
   通过做一些你喜欢做的事来减压。
   ____.
4. The more a strain of bacteria is exposed to an antibiotic, the more likely it is to become resistant.
   如果你把皮肤暴露在阳光下，皮肤会晒伤的。
   ____.
5. When you have a large-scale data set, the human mind can't cope with that, for it's too complex.
   不论情况多么复杂，他都能应付自如。
   ____.

Passage 2

Why Make a Diagnosis?

A cow lies in the middle of the pasture after the herd has moved in to be milked. The owner looks closely at the cow and urges her to get up. She makes a feeble attempt to rise but is either unable or too depressed to try. Is the cow injured? Has she become intoxicated by something she ate? Does she have a severe systemic infection? Is there any functional change in her nervous system, digestive system, or urogenital system? The process of determining the nature of disease is diagnosis. Making a diagnosis is a systematic process of collecting all the facts and making an objective evaluation of them while considering all that is known about specific diseases and their signs.

Why make a diagnosis? In the example of the"downer" cow, a diagnosis is necessary for the protection of the rest of the herd. If the cause of her disease is available to the rest of the herd, it must be eliminated to prevent further loss. Of secondary importance is the correct approach to restoring the cow to health, which depends on determining the nature of the disease. Most diagnosticians have learned not to jump to conclusions, even when apparently clear-cut signs point to specific diseases.

New Words

diagnosis	[ˌdaɪəgˈnəʊsɪs]	*n.*	诊断
herd	[hɜ:d]	*n.*	牧群
feeble	[ˈfi:bl]	*adj.*	虚弱的
intoxicated	[ɪnˈtɒksɪkeɪtɪd]	*adj.*	中毒的
systemic	[sɪˈsti:mɪk]	*adj.*	系统的
infection	[ɪnˈfekʃn]	*n.*	传染
nervous	[ˈnɜ:vəs]	*adj.*	神经系统的
urogenital	[ˌjʊərə(ʊ)ˈdʒenɪt(ə)l]	*adj.*	泌尿生殖器的
objective	[əbˈdʒektɪv]	*adj.*	客观的
evaluation	[ɪˌvæljuˈeɪʃn]	*n.*	评价
eliminate	[ɪˈlɪmɪneɪt]	*v.*	清除
secondary	[ˈsekəndri]	*adj.*	次要的
restore	[rɪˈstɔ:(r)]	*v.*	恢复
diagnostician	[ˌdaɪəgnɒsˈtɪʃən]	*n.*	诊断专家

Phrases & Expressions

jump to conclusions	妄下结论
nervous system	神经系统
digestive system	消化系统
urogenital system	泌尿生殖系统
clear-cut	明显的

Check your understanding

Choose the correct answers according to the information given in the passage.

1. The process of determining the ________ is diagnosis.

A. action of disease B. name of disease C. nature of disease

2. Making a diagnosis is a systematic process of collecting all the facts and making a/an ________ of them.

   A. subjective evaluation B. objective evaluation C. rough evaluation

3. According to the passage, it is important to make a diagnosis as soon as possible because ________.

   A. it is necessary for the protection of the rest of the farm animals

   B. it is the right method to make the cow become healthy again

   C. both A and B

4. What does "diagnosticians" mean in this passage?

   A. A person who is qualified to prepare and sell medicines.

   B. A specialist or expert in making diagnoses.

   C. An animal caregiver.

5. What happens if an animal disease is not diagnosed? The following statements are correct EXCEPT ________.

   A. It will cause animal death

   B. It will lead to further loss

   C. The rest of the animals will definitely be infected

Task 4 Related Information

Understand the nutritional information about eggs.

Time and time again, the differences between cage-raised and free-range eggs have been apparent. Free-range eggs may contain:

- $^{1}/_{3}$ less cholesterol
- $^{1}/_{4}$ less saturated fat
- $^{2}/_{3}$ more vitamin A
- 2 times more omega-3
- 3 times more vitamin E
- 7 times more beta-carotene

Egg Nutrition

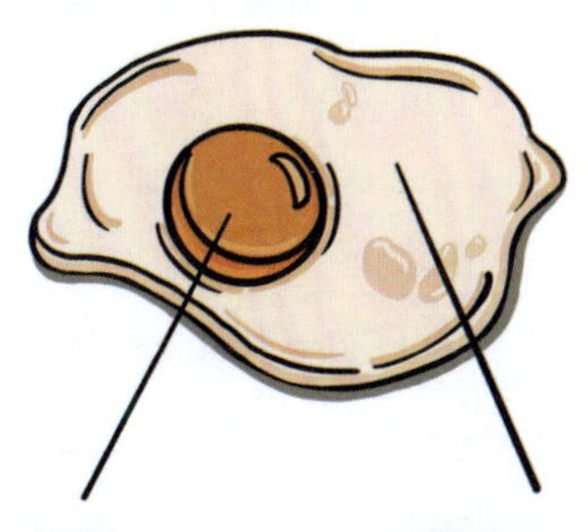

Yolk		White	
Fat	**4.5 g**	Fat	0 g
Sat. Fat	**1.6 g**	Sat. Fat	0 g
Cholesterol	**184 mg**	Cholesterol	0 mg
Carbohydrates	0.5 g	Carbohydrates	0 g
Protein	2.5 g	**Protein**	**4 g**

Not only do these eggs clearly win in the nutrition category, but free-range eggs are 98 percent less likely to carry salmonella! It's no surprise, considering how cage-raised hens are expected to live.

Therefore, free-range eggs are a much better option to ensure food safety, support ethical farming practices and maximize the potential egg nutrition facts.

Task 5 Pop Quiz

Ⅰ. Working with words

Match the following English technical terms with their Chinese equivalents.

A. vaccine B. animal welfare

C. veterinarian D. veterinary drug residues

E. bird flu（avian influenza） F. virus

G. surgical operation H. contagious viral disease

I. rational drug therapy J. foot-and-mouth disease

K. feed management L. animal husbandry

1. 禽流感 （ ）	6. 口蹄疫 （ ）
2. 外科手术（ ）	7. 合理药物治疗（ ）
3. 兽医 （ ）	8. 畜牧业 （ ）
4. 兽药残留（ ）	9. 疫苗 （ ）
5. 饲料管理（ ）	10. 动物福利 （ ）

Ⅱ. Multiple choice

1. ________ plays a vital role in the control of disease transmission from animals to humans.

A. Animal welfare B. Veterinary drug C. Animal health

2. ________ is the largest cost item in livestock and poultry production, accounting for 60%-70% of total expenses.

A. Electricity B. Water C. Animal feed

3. The most common and widely raised poultry birds are ________.

A. chickens B. turkeys C. quails

4. The FMD virus causes illness in cows, pigs, sheep, goats, deer, and other animals with divided hooves. It does not ________ horses, dogs, or cats.

A. lead B. cause C. affect

5. Veterinary drug ________ are the very small amounts of veterinary drugs that can remain in animal products and therefore make their way into the food chain.

A. leaves B. residues C. items

6. Eggs are also a great source of ________ and can help provide a wealth of important nutrients.

A. protein B. sugar C. salt

7. Animal ________ is the physical and psychological well-being of animals.

   A. breeding   B. management   C. welfare

8. Each goat is equipped with an electronic ________ tag, which identifies the goat as soon as it enters its space on the rotary milking machine.

   A. intelligence   B. identification   C. effect

Ⅲ. Fill in the blanks with the proper choices given.

___(1)___, causing severe economic losses for many pig farmers and pork producers. In addition, the number of ASF cases in wild boar populations has dramatically increased over the past few years. Evidence supports direct contact with infectious domestic pigs and wild boars, and consumption of contaminated feed, ___(2)___. However, significant knowledge gaps highlight the urgent need for research to investigate the dynamics of indirect transmission via the environment, the minimal infective doses for contaminated feed ingestion, the probability of effective contacts between infectious wild boars and domestic pigs, the potential for recovered animals to become carriers and a reservoir for transmission, the potential virus persistence within wild boar populations and the influence of human behaviour for the spread of ASFV. This will provide an improved scientific basis to optimize current interventions and develop new tools and strategies to___(3)___.

With high-virulence forms of the virus, ___(4)___:

- ___(5)___
- loss of appetite
- haemorrhages in the skin and internal organs
- death in 2-10 days on average

A. reduce the risk of ASFV transmission to domestic pigs

B. ASF is characterized by

C. as the main transmission routes of ASF virus（ASFV）to domestic pigs

D. high fever

E. African swine fever (ASF) is a major threat to the pig industry in Europe

UNIT 10 Environmental Planning

Task 1 First Sight

1. Match the following pictures with the key words given.

A. flower arrangement

B. landscape plan

C. rooftop garden

D. residential landscaping

(1)

(2)

(3)

(4)

2. Answer the question by matching the following items.

What do landscape architects do?

□1) meet with	A. landscaping materials
□2) select	B. graphic representations of plans
□3) inspect	C. environmental reports on land conditions
□4) prepare	D. clients
□5) analyze	E. project progress

Task 2 Better Acquaintance

Conversation

At the Florist's

Florist: Morning, madam. What can I do for you?

Nancy: I want to book a bouquet of flowers for Mother's Day. What do you recommend?

Florist: Carnations. They are the perfect gift for mums on Mother's Day. And roses are also used to thank mothers for their love and care.

Nancy: I see. A bouquet of roses, please. By the way, I also want some flowers to decorate my living room. Could you give me some advice?

Florist: No problem. Lilies and sunflowers are the most common choices. Lily is popular for its unique fragrance, and sunflower can brighten up your living room and give you a welcoming look.

Nancy: I want both. Could you tell me how to choose cut flowers?

Florist: Choose the ones with buds that are just starting to open over those that are already fully open. The flower buds will continue to open if kept in a vase with water.

Nancy: Thank you. Is there anything I can do to prolong the vase life of fresh flowers?

Florist: Of course. There are some easy things you can do to lengthen their vase life. First, remove the leaves that will be below the water line. If you leave them there, they'll start to foul the water and your flowers will die sooner. Second, cut an inch or two from the stem at an angle, and this will increase the surface area the flower can use to absorb water.

Nancy: That sounds easy. I will do as you say. And how often should I change the water?

Florist: You'd better replace the water every day. If you can't, change the water at least every couple of days and trim the stems at the same time.

Nancy: Got it. It's very kind of you.

New Words

florist	[ˈflɒrɪst]	*n.*	花艺师
bouquet	[buˈkeɪ]	*n.*	花束
carnation	[kɑːˈneɪʃn]	*n.*	康乃馨
lily	[ˈlɪli]	*n.*	百合花
fragrance	[ˈfreɪgrəns]	*n.*	香味
bud	[bʌd]	*n.*	花苞
foul	[faʊl]	*v.*	弄脏；污染
stem	[stem]	*n.*	（花草的）茎
absorb	[əbˈzɔːb]	*v.*	吸收

Phrases & Expressions

brighten up	为……增辉添彩
at an angle	斜的

Projects

Ⅰ. Substitution drills

1. A green roof with a living system of plants and soil can actually

在夏季的时候降低温度.
节省冬季供暖费用.
清洁并储存雨水.
为昆虫和鸟类提供栖息地.
减少城市热岛效应.

2. Flower arrangement can produce a visually pleasing display of

鲜切花.
干花.
仿真花.
野花.
绢花.

3. A well-designed front yard will

有助于提高宅院的外观魅力.
是放松和娱乐的好场所.
每当回到家就给你一种快乐的感觉.
创造一种宁静的氛围.
是孩子玩耍的空间.

4. A professional landscape design team can create a/an outdoor space for your front yard.

易于维护的
亲近自然的
低成本的
美观的
舒适而安全的

5. The main disciplines within landscape architecture are

景观规划.
景观设计.
景观管理.
城市设计.
社区规划.

Ⅱ. The following are questions and answers about environmental planning between two friends. Join the questions 1-8 to the answers A-H. Then act out some of the mini-dialogs with your partner.

1. Could you tell me the features of the playground you are designing? ________
2. How much does it cost to get a landscape design? ________
3. I want to design the front yard on my own. Could you introduce some Apps for landscape design to me? ________
4. It's said that cut flowers can't be placed near fruit. Is that true? ________
5. What in-season flowers can I choose to beautify my house? ________
6. Which is better for my front yard, native plants or the uncommon ones in my place? ________
7. What does a landscape architect do? ________
8. What shall we do with grass clippings, throw out as yard waste or recycle them by returning them to the lawn? ________

A. It is better and smarter to choose native plants, as these kinds of plants have adapted to the natural habitat and it will be easier for you to grow them.

B. Yes, it is. The reason is that fruit releases ethylene gas that will speed up the maturation process and cause wilting.

C. As a general rule, grass clippings of an inch or less in length can be left on the lawn where they will filter down to the soil surfaces and decompose quickly. And remove longer clippings.

D. They design parks, campuses, streetscapes, trails, plazas, and other projects that help define a community.

E. The playground is a child-oriented landscape that connects nature to a child's sense exploration, adventure and fantasy.

F. The most popular flowers in summertime include lilies, gerbera daisies and sunflowers.

G. The price can vary greatly depending on the size and scope of the project.

H. Certainly. Landscaper's Companion and Home Outside are popular Apps. They offer services from phone/video consultations with professional landscape designers to complete designs.

Ⅲ. Fill in the blanks in the conversation by translating the Chinese into appropriate English.

Woman: Hello sir. May I help you?

Men: I'd like to（1）______________________________（买些鲜花送给妻子）.

Woman: No problem. May I ask（2）_________________________（是什么场合）?

Men: It's our 10th anniversary.

Woman: I see. How about this arrangement? It has（3）___________（两打长茎红玫瑰）. Traditionally, the rose（4）________________________（被认为是爱之花）.

Men: But my wife doesn't like roses. What else do you recommend?

Woman: In that case, I recommend this sweet bouquet, bright colors and beautiful flowers. It is（5）______________（一款经典的结婚纪念日插花）.

Men: Oh, it contains orange and pink lilies, my wife's favourite flowers. OK, I'll take it.

Ⅳ. Remember the useful expressions concerning garden design.

1. Use warm-colored plants, like those that are red or orange, to add warmth and excitement to the garden.
2. Up-lighting is quite effective in drawing attention to an area wanted to be highlighted.
3. Ornamental trees can be used to create a focal point in the garden.
4. A patio or a deck is an essential element in today's backyards.
5. The easiest way to create shade on the patio or in the backyard is to add a patio umbrella.
6. Evergreens can provide year-round color and texture in a landscape.
7. A deck with pergola in the garden will provide you a good place for rest and entertainment.
8. Flagstone walkways are durable and don't require much maintenance.

Task 3 Further Development

Passage 1

Pre-reading Task

Read the following statements and tick in the columns under "True", "False" or "Unsure".

Statements	Ture	False	Unsure
Landscape designers should understand the clients' wishes and try to meet their needs.			
The business of a landscape architectural firm ranges from residential design to commercial design.			
A landscape architectural firm should provide on-site supervision during the contractor's installation.			
The firm has to offer a preliminary plan to their clients before the initial meeting with them.			
A landscape architectural firm has to prepare cost estimates for the clients.			

Introduction to a Landscape Architectural Firm

We are a landscape architectural firm that listens to and works closely with our clients to translate their goals and wishes into a design that is creative, cost-effective and suited to their unique lifestyle.

Our strength lies in our ability to arrange all of the separate elements of our clients' ideal outdoor living space into an environment that is enjoyable, safe and dynamically changing through the seasons.

Our landscape design begins with a preliminary plan showing the elements discussed in the initial meeting with the client. Once this plan is approved, the process proceeds to the complete working drawings which include: construction drawings, lighting, drainage, planting and irrigation plans and complete details.

The development of all details during the design phase ensures that landscape contractors will present comparable, competitive bids and the problems and unwanted surprises during actual construction will be reduced or eliminated.

We also offer on-the-job supervision to ensure coordination of the plans with the contractor's installation.

We are your best choice for landscape architectural design because:

We bring over 25 years' experience in the landscape industry including large and small residential and commercial design.

Our extensive experience includes creative planting design.

We are specialists in drought tolerant and water-saving gardens.

We offer all styles of landscape design, including:

Theme gardens（European, Asian, California, etc.）

Formal gardens

Natural-style gardens

Herb gardens

Water-themed gardens

New Words

landscape	[ˈlændskeɪp]	*n.*	景观
architectural	[ˌɑːkɪˈtektʃərəl]	*adj.*	建筑方面的
client	[ˈklaɪənt]	*n.*	客户
element	[ˈelɪmənt]	*n.*	基本部分
dynamically	[daɪˈnæmɪkəli]	*adv.*	动态地
preliminary	[prɪˈlɪmɪnəri]	*adj.*	初步的
initial	[ɪˈnɪʃl]	*adj.*	最初的
approve	[əˈpruːv]	*v.*	同意
proceed	[prəˈsiːd]	*v.*	接着做
phase	[feɪz]	*n.*	阶段
contractor	[kənˈtræktə(r)]	*n.*	承包商
bid	[bɪd]	*v.*	出价
supervision	[ˌsuːpəˈvɪʒn]	*n.*	监督
installation	[ˌɪnstəˈleɪʃn]	*n.*	安装
residential	[ˌreziˈdenʃl]	*adj.*	住宅的
extensive	[ɪkˈstensɪv]	*adj.*	大量的
drought	[draʊt]	*n.*	干旱
theme	[θiːm]	*n.*	主题

Phrases & Expressions

translate...into...	把（思想、感情等）用另一种形式表达出来
lie in	在于……
drought tolerant and water-saving garden	抗旱节水型花园
herb garden	草本园
natural-style garden	自然风格花园
water-themed garden	水系主题花园

Projects

Ⅰ. Fill in the blanks with the proper words given, changing the form if necessary.

installation	irrigation	supervision	eliminate
effective	arrange	residential	commercial

1. They decide to __________ a lighting system in the front yard.
2. They are the leading manufacturer in both defence and __________ products.
3. The police have __________ two suspects from their investigation.
4. Services need to be more __________ organized than they are at present.
5. 10 Downing Street is the British Prime Minister's official __________.
6. Children should not be left to play without __________ of adults.
7. Her flower __________ won first prize.
8. They __________ their crops with water from this river.

Ⅱ. Translate the following sentences by imitating the examples given, paying attention to the underlined phrases or structures.

1. Our strength lies in our ability to arrange all of the separate elements of our clients' ideal outdoor living space into an environment that is enjoyable, safe and dynamically changing through the seasons.
   主要困难在于找到一个好时机。
   ____.
2. We offer all styles of landscape design.
   他们决定把这份工作给乔。
   ____.
3. Our landscape design begins with a preliminary plan showing the elements discussed in the initial meeting with the client.
   政府决定重修在地震中被毁坏的大桥。
   ____.
4. We also offer on-the-job supervision to ensure coordination of the plans with the contractor's installation.
   那本书确定了他的成功。
   ____.
5. Once this plan is approved, the process proceeds to the complete working drawings.
   让我们转入下一个话题的讨论吧。
   ____.

Passage 2

How to Arrange Flowers

Flowers can brighten an interior space and play traditional roles in major events like weddings. With a little planning and a little know-how, you can create a lovely floral arrangement for any occasion.

First: Planning Your Flower Arrangement

1. Use complementary colors. Color is very important when it comes to flower arrangement, but there are no hard and fast rules when it comes to what color combinations work best. It will depend on the style and mood you're going for.

2. Consider the location of the finished arrangement. Consider the color scheme and general mood of the room you intend to place it in, so you can choose a design to match. You will also need to consider the amount of space available.

Second: Making the Arrangement

1. When you begin working on the arrangement, start with the largest or most prominent variety of flowers.

2. Layer the flowers as you go. Once you have finished with the first circle of flowers, move on to the second, using a different variety of flowers. Try to create a domed effect by leaving the stems slightly longer on the inner flowers. The finished arrangement should look like a bunch of flowers growing on a hilltop.

3. One of the main rules when it comes to flower arranging is to use an odd number of each variety of flowers.

4. Pay attention to height and width. The general rule when it comes to height is that your arrangement should be one and half times the height of the vase or container it's held in. There's no clear-cut rule when it comes to the width of your arrangement, but it should be wide enough to balance out the height.

5. When you have arranged all of the flowers to your liking, you can add the final touches by inserting any greenery, leaves, berries or other decorations.

New Words

interior	[ɪnˈtɪəriə(r)]	*adj.*	内部的
arrangement	[əˈreɪndʒmənt]	*n.*	布置
complementary	[ˌkɒmplɪˈmentri]	*adj.*	相称的；互补的
prominent	[ˈprɒmɪnənt]	*adj.*	重要的
greenery	[ˈgriːnəri]	*n.*	青枝绿叶
berry	[ˈberi]	*n.*	浆果
touch	[tʌtʃ]	*n.*	细微之处
decoration	[ˌdekəˈreɪʃn]	*n.*	装饰品

Phrases & Expressions

play a role in	在……中起作用
balance out	（使）平衡；（使）相抵
pay attention to	重视；对……注意

Check your understanding

Decide whether the following statements are true or false according to the information given in the passage.

1. The color theme and the size of a flower arrangement should match the color and mood of the area where it is placed.
2. Place the largest or more dominant flowers in the middle as the first layer when you make a flower arrangement.
3. Each layer of the arrangement must use the same variety of flowers.
4. Consider the colors carefully when choosing flowers for an arrangement since there are hard and strict rules for color combinations.
5. The last step of an arrangement is adding some leaves, stalks or other decorations to it.

Task 4 Related Information

Features of Traditional Chinese Gardens

Chinese gardens are constructed to recreate and miniaturize larger natural landscapes. Traditionally, Chinese gardens blend unique, ornate buildings with natural elements.

Almost every Chinese garden contains architecture, like a building or pavilion, decorative rocks and a rock garden, plants, trees and flowers, and water elements, like ponds. Most Chinese gardens are enclosed by a wall and some have winding paths.

Chinese gardens aren't just thrown together. Instead, they're deliberately designed and visitors should walk through them in the particular order that the garden was laid out.

Task 5 Pop Quiz

Ⅰ. Working with words

Match the following English technical terms with their Chinese equivalents.

A. feature paving

B. needle-leave plant

C. pattern dwarf hedge

D. clump planting

E. feature wall

F. clipped hedge

G. hand-draw perspective

H. construction drawing

I. broad-leave plant

J. opposite planting

K. flower bed

L. nursery

1. 丛植 (　　)	6. 苗圃 (　　)
2. 特色景墙 (　　)	7. 针叶植物 (　　)
3. 整剪绿篱 (　　)	8. 特色铺装 (　　)
4. 手绘透视图 (　　)	9. 阔叶植物 (　　)
5. 图案矮篱 (　　)	10. 对植 (　　)

Ⅱ. Multiple choice

1. Flowers that are not available naturally, but made artificially from various materials are known as ________ .

A. artificial flowers　　B. dried flowers　　C. cut flowers

2. When choosing evergreen trees and shrubs for landscaping, you should choose plants with ________ that match the conditions of your landscape.

A. colors　　B. shapes　　C. growing requirements

3. ________ is the design of outdoor public areas, landmarks, structures to achieve environmental, social-behavioral or aesthetic outcomes.

A. Flower arrangement　　B. Landscape architecture　　C. Interior design

4. A flower frog is a heavy object that is placed in the bottom of flowers vases to hold the flower stems in place, and it is also called pin frogs. The underlined words mean ________.

A. 剑山/花插　　B. 花泥　　C. 花土

5. Floral ________ are always used to increase the lifespan of fresh cut flowers.

A. fertilizers　　B. preservatives　　C. artificial flavors

6. With the use of decorative edging, it is easier to create solid boundaries between your lawn and garden, and also maintain a freshly groomed look at all times. The underlined words mean________.

A. 围墙　　B. 栅栏　　C. 饰边

7. ________ are the hardworking and creative people who draw and then execute plans for public parks, green spaces, businesses and private residences.

A. Florists　　B. Landscape architects　　C. Engineers

8. As the name suggests, ________ are used to decorate tables, be it the dining or the centre table. So the size and the height of the arrangement have to be proportionate to the table.

A. centrepiece arrangements　　B. basket arrangements　　C. wreath arrangements

Ⅲ. Fill in the blanks with the proper choices given.

The tricks to choosing plants for landscaping your garden are as follows:

Decide on a color scheme.

When you are putting together the plan for your garden, color is one of the first things to decide on. Just like when you are decorating a room,___(1)___.

Consider all seasons.

Remember that even though your garden may not be flowering all year round, you can still see it all year round. Make sure ___(2)___.

Where is the sun?

To choose plants for landscaping that will grow well, you will need to take note of which areas of your garden get the most sun and which get the least. Make sure to use plants that ___(3)___.

Check the zone.

This is one of the most important steps for choosing plants for landscaping your garden. Make sure that the plants you buy are meant ___(4)___.

How much maintenance are you willing to do?

If you are trying to minimize the amount of work to be done in the yard, make sure to do a little research on the plants you are planning to buy to see ___(5)___.

Keeping all of these plant selection tips in mind will help you make better plant choices for landscaping your garden and keep it as low maintenance as possible.

A. what their maintenance requirements are

B. to grow in the gardening zone where you live

C. the color scheme will set the tone for the space

D. to plant things that will add winter interest

E. need the amount of light provided

UNIT 11 Applying for a Job

Task 1 First Sight

1. Match the following pictures with the key words given.

A. want ad B. career fair C. job interview D. resume/CV

(1)

(2)

(3)

(4)

2. Answer the question by matching the following items.

What are the key factors for a successful job hunting?

□1) sophisticated	A. educational background
□2) proper	B. attitude and enthusiasm
□3) solid	C. interview clothing
□4) diverse	D. qualifications and certificates
□5) positive	E. social skills

Task 2 Better Acquaintance

Conversation

A Job Interview

(Mary Smith is being interviewed. She is applying for the position of a typist.)

Mary: Good morning, sir. My name is Mary Smith. I'm coming to apply for the position of typist in your company.

Interviewer: Glad to meet you. Have a seat please. Let me ask you a few questions. Can you type fast?

Mary: Yes. I can type about 100 words a minute.

Interviewer: Good. And what foreign languages do you know?

Mary: I'm fluent in both English and French.

Interviewer: Really? That's great! We need a typist who knows both English and French. Are you willing to work overtime at weekends?

Mary: Work overtime at weekends? Oh, I will accept it if it's necessary.

Interviewer: What's your expected salary?

Mary: 5,000 yuan will be fine.

Interviewer: By the way, where do you live? Do you live near our company?

Mary: Yes, it's about twenty minutes' walk from here.

Interviewer: We are very much satisfied with you, Miss Smith. You can start to work tomorrow.

Mary: You mean I can have the job? Oh, thank you so much. I'll try my best to do it well.

New Words

interview	[ˈɪntəvjuː]	*n.*	面试

apply	[ə'plaɪ]	*v.*	申请
typist	['taɪpɪst]	*n.*	打字员
fluent	['flu:ənt]	*adj.*	流利的
salary	['sæləri]	*n.*	薪金
satisfied	['sætɪsfaɪd]	*adj.*	满意的

Phrases & Expressions

apply for	申请
try one's best	尽力

Projects

Ⅰ. Substitution drills

1.What kind of work do you want to do?

I've always been interested in

市场营销.
广告业.
办公室工作.
IT行业.
在大学教书.

2. What are you going to do for a living after graduation from university?

I'm going to be

自由撰稿人.
秘书.
软件工程师.
公务员.
导游.

3. You are working in this company now, aren't you?

Yes, I've been working here for two years as

总经理助理.
营销经理.
首席财务官.
首席信息官.
人力资源经理.

4. Why are you interested in our company?

Because your company

提供丰厚的薪水.
拥有良好的声誉.
拥有很好的销售记录.
给员工提供晋升机会.
是有名的软件公司.

5. How do you like the work here?

It's very

有趣.
具有挑战性.
无聊.
困难.
危险.

Ⅱ. The following are questions and answers about job hunting between two people. Join the questions 1-8 to the answers A-H. Then act out some of the mini-dialogs with your partner.

1. Did you have any trouble finding us? ________
2. What kind of person do you think you are? ________
3. What did you enjoy most about your last job? ________
4. How do I know if there is a vacancy in a company? ________
5. Where do you see yourself in five years? ________
6. How are you getting on with your studies? ________
7. Have you found another job yet, Sally? ________
8. What am I going to wear for the interview? ________

A. No, I haven't. It's very hard for college graduates to find an ideal job in big cities.

B. It's easy. You can view the homepage of a company, and see if it's in need of staff.

C. You may wear your favorite suit with a tie.

D. Well, I'd like to work for an organization where I can build a career. I'm eager to assume more management responsibilities and get involved in product strategy.

E. Well, I am energetic and outgoing. Generally speaking, I am an open-minded person.

F. No, it isn't too difficult to find the office.

G. I loved interacting with customers and it was very exciting to deal with people from other cultures.

H. I'm doing well at school.

Ⅲ. Fill in the blanks in the conversation by translating the Chinese into English.

A Job Offer

(Tom has just accepted an offer. Now he is talking with David about his new job.)

David: Hi, Tom, (1) ____________________ (你接到工作邀请了吗)?

Tom: Yes, I have. I will start working in two weeks and the salary is good.

David: Really? Congratulations. What company are you going to work for?

Tom: (2) ____________________________________ (是一家网站设计公司).

David: Does the job require a lot of overtime and even weekends?

Tom: Yes, I nearly have no free time to find a girlfriend. But I think it's very important for newcomers to get their feet planted firmly on the ground, so I accepted the job.

David: Wow, you're awesome. By the way, (3) ______________________________ (从你家到公司有多远)?

Tom: The company is in the downtown area and (4) ______________________________ (大约半个小时的车程).

David: Good. And you can also go to work by subway.

Tom: Yes, it's really convenient.

David: (5) ______________________ (祝你工作愉快).

Tom: Thank you.

Ⅳ. Remember the useful expressions concerning job hunting.

1. Now we have some questions about your work experiences. Could you tell us a little about yourself?
2. Why do you want to come to work with us?
3. Why are you leaving your current job?
4. What is your expected salary?
5. I am interested in the job position and the attractive salary you offer.
6. Would you mind traveling on business?
7. We'll inform you the final decision within two weeks.
8. What do you see as your strengths and weaknesses?

Task 3 Further Development

Passage 1

Pre-reading Task

Read the following statements and tick in the columns under "True", "False" or "Unsure".

Statements	True	False	Unsure
A formal business suit is always the right outfit for an interview.			
Reference letters can make or break a job search.			
When applying for a job, you have to figure out a way to differentiate yourself.			
It's always important to check your resume to make sure there are no mistakes in grammar.			
The cover letter should be targeted to the position you are applying for.			

7 Tips for Applying for Your Next Job

Whether you are looking for a part-time job, a job in your current industry or thinking about changing careers, you have to figure out a way to differentiate yourself. There are some general rules that applicants should follow if they want to give themselves a fighting chance to get an interview.

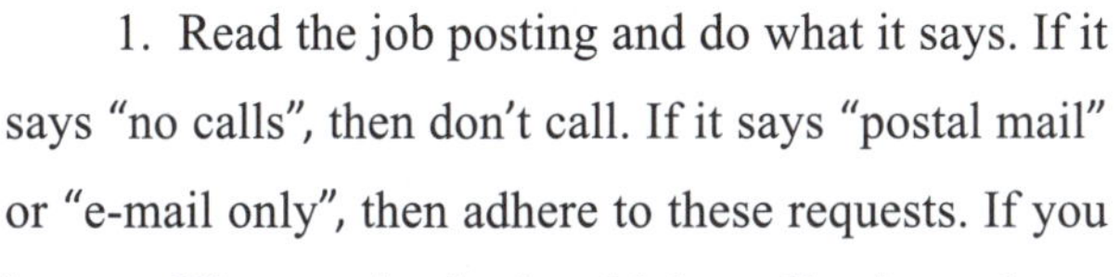

1. Read the job posting and do what it says. If it says "no calls", then don't call. If it says "postal mail" or "e-mail only", then adhere to these requests. If you have to fill out an institutional job application to be considered, then do it.

2. Personalize your cover letter. Don't just send a general cover letter. It doesn't need to be lengthy. You should get right to the point as to why you should be considered. Make it appear that this is the one job you want and you have written the letter specifically for this job.

3. Be organized, attractive and grammatically correct in your presentation. Mistakes or sloppiness is almost always a recipe to have your materials set aside. Color is not necessarily a bad thing as long as it is not overdone.

4. Don't chase unicorns. If you are a middle school coach or never coached in college, then don't apply for a college job. You are not going to have a chance.

5. Don't wait. It doesn't matter when the deadline is. Just get your materials in ASAP. Many places begin to interview before the deadline.

6. Know what the school or person in charge of the hiring is looking for. Sometimes, it takes a little internet research or keeping your ears open.

7. If you have some top-notch reference letters, then send them with your materials. This is especially helpful if they are heavy hitters.

New Words

current	[ˈkʌrənt]	*adj.*	现在的
differentiate	[ˌdɪfəˈrenʃieɪt]	*v.*	区分
applicant	[ˈæplɪkənt]	*n.*	申请人
adhere	[ədˈhɪə(r)]	*v.*	遵守
institutional	[ˌɪnstɪˈtjuːʃənl]	*adj.*	机构的
personalize	[ˈpɜːsənəlaɪz]	*v.*	使个性化

lengthy	[ˈleŋθi]	*adj.*	冗长的
specifically	[spəˈsɪfɪkli]	*adv.*	具体地；特别地
grammatically	[grəˈmætɪklɪ]	*adv.*	文法上地；语法上地
presentation	[ˌpreznˈteɪʃn]	*n.*	提交；报告
sloppiness	[ˈslɒpɪnəs]	*n.*	马虎
recipe	[ˈresəpi]	*n.*	秘诀
unicorn	[ˈjuːnɪkɔːn]	*n.*	（传说中的）独角兽
deadline	[ˈdedlaɪn]	*n.*	截止日期
top-notch	[ˌtɒp ˈnɒtʃ]	*adj.*	卓越的
reference	[ˈrefrəns]	*n.*	证明信；证明人
hitter	[ˈhɪtə(r)]	*n.*	要员；强有力的人

Phrases & Expressions

fill out	填写
adhere to	遵循
in charge of	掌管
figure out	想出
set aside	把……放置一旁
ASAP（as soon as possible）	尽快

Projects

Ⅰ. Fill in the blanks with the proper words and expressions given, changing the form if necessary.

apply	differentiate	length	attract
personalize	request	sloppy	interview

1. It creates a very bad impression if you're late for an ________.
2. The members in the testing team were quite flexible and could change their schedule upon ________.
3. It is wrong to ________ students according to their family backgrounds.
4. Times have changed and so have the criteria（标准）for resume ________.
5. The ________ should be flexible, creative, and be able to work in a team atmosphere.
6. I'm worried that someone more ________ than me will get the job.
7. If your work is ________ or if you make a lot of mistakes, you will have to change your ways.

8. Using social media to learn more about guests will help the hotel to add some touches and details that ________ the guest experience.

Ⅱ. Translate the following sentences by imitating the examples given, paying attention to the underlined phrases or structures.

1. Color is not necessarily a bad thing as long as it is not overdone.
只要准备充分，你就会在面试时充满信心。
____.

2. Know what the school or person in charge of the hiring is looking for.
我不在的时候，张先生是办公室负责人。
____.

3. You should get right to the point as to why you should be considered.
至于他究竟是否适合这份工作，他实在拿不准。
____.

4. Mistakes or sloppiness is almost always a recipe to have your materials set aside.
如果他够幸运的话，他的商业计划会被接受。
____.

5. You have to figure out a way to differentiate yourself.
当一种新疾病暴发时，专家们必须以最快的方式弄明白该干什么。
____.

Passage 2

Tips for Making the Best Impression at a Job Interview

An interview is not only meant to assess your skills but also your personality. Therefore, whether you are going to be considered for the job or not is determined by the impression you create on the interviewers. Here are some suggestions:

1. Dress Decently

The type of dress or clothes you put on says so much about you. Do not dress in casual or dirty clothes and make sure that you are groomed properly.

2. Arrive on Time

Punctuality is a quality that employers will be looking out for. To prevent this from ever happening, familiarize yourself with the venue and determine how much time you'll need to get there. You can go to bed early the night before the interview so that you can wake up on time the next morning.

3. Be Confident

During the interviewing process, just be yourself. To create a good impression, show that you're interested and enthusiastic about the job. Convince the interviewer that you are the right person for the job by highlighting your accomplishments when working in your previous job or emphasizing your achievements both in academics and extracurricular activities if you are a student.

4. Ask Smart Questions

In most cases, you will be given a chance to ask the interviewers a few questions at the end of the interview. Given a chance, ask intelligent and relevant questions that can show your deep interest in the company or organization.

5. Show Appreciation

Before you leave after the interviewing process, do not forget to thank your interviewers by giving them a firm handshake and saying kind words such as "Thank you," "I am much obliged," or "I would be glad to hear from you again."

These tips on how to make a good first impression at a job interview cannot guarantee that you will get the job, but they can certainly get you a little closer.

New Words

impression	[ɪmˈpreʃn]	*n.*	印象
assess	[əˈses]	*v.*	评估
personality	[ˌpɜːsəˈnæləti]	*n.*	性格
decently	['diːsntlɪ]	*adv.*	体面地
casual	[ˈkæʒuəl]	*adj.*	非正式的
groom	[gruːm]	*v.*	打扮
punctuality	[ˌpʌŋktjʊˈælɪti]	*n.*	准时
familiarize	[fəˈmɪliəraɪz]	*v.*	使熟悉
venue	[ˈvenjuː]	*n.*	聚会地点
convince	[kənˈvɪns]	*v.*	使信服
highlight	[ˈhaɪlaɪt]	*v.*	强调
accomplishment	[əˈkʌmplɪʃmənt]	*n.*	成就
emphasize	[ˈemfəsaɪz]	*v.*	强调
achievement	[əˈtʃiːvmənt]	*n.*	成就
academics	[ˌækəˈdemɪks]	*n.*	学术知识
extracurricular	[ˌekstrəkə'rɪkjʊlə(r)]	*adj.*	课外的
obliged	[əˈblaɪdʒd]	*adj.*	感激的
guarantee	[ˌgærənˈtiː]	*v.*	保证

Phrases & Expressions

look out for　　留意

prevent...from...	阻止做某事
in most cases	大多数情况下
familiarize...with...	使某人熟悉某物
wake up	醒来
be enthusiastic about	对……热心

Check your understanding

Choose the correct answers according to the information given in the passage.

1. What kind of people is the text mainly intended for? ________.
   A. Interviewers
   B. Job-hunting people
   C. Employers
2. According to the passage, which of the following statements on interviews is right? ________.
   A. Pretend to be someone you are not
   B. Wear new clothes for the interview
   C. Show up for the interview on time
3. This article is about ________.
   A. how to make a great first impression at a job interview
   B. how to get a job interview
   C. how to answer questions at a job interview
4. Which of the following statements is mentioned in the passage? ________.
   A. Tell interviewers all about yourself whether it is questioned or not
   B. Ask an experienced friend to go along with you if you feel nervous
   C. Show your great enthusiasm for the job and the company
5. It can be learned from the passage that ________.
   A. the applicant should ask for business cards at the end of the interview
   B. it matters much for the applicant to leave a good first impression on the interviewers
   C. a good first impression at a job interview guarantees a job

Task 4 Related Information

Proper Dress for a Job Interview

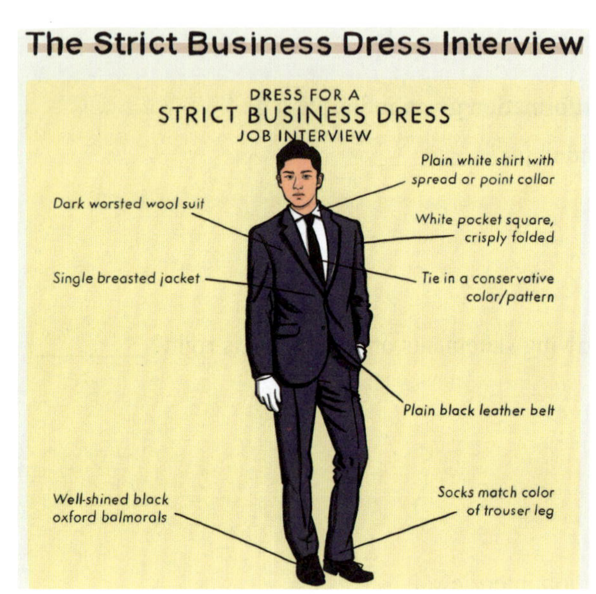

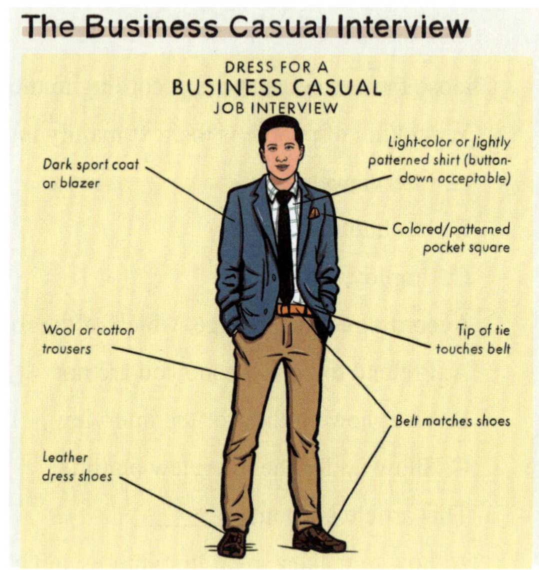

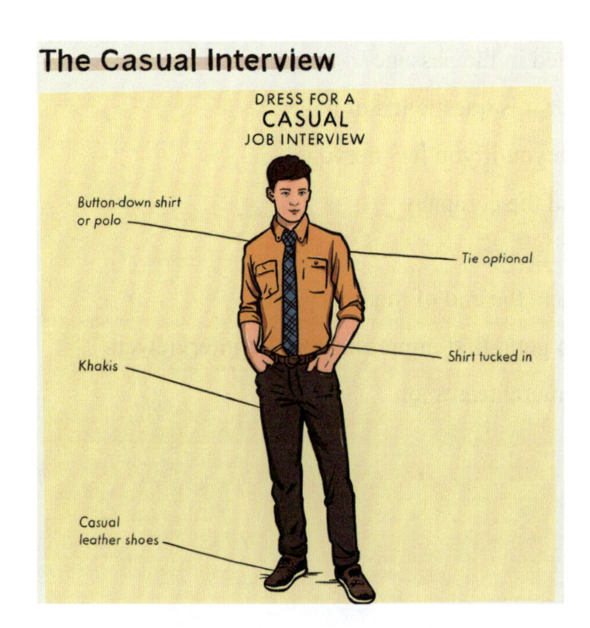

Task 5 Pop Quiz

Ⅰ. Working with words

Match the following English words and phrases with their Chinese equivalents.

A. skills and strengths
B. transcript
C. internship
D. work experiences
E. reference
F. honors and awards
G. educational background
H. job objective
I. English proficiency
J. resume
K. personal data
L. cover letter

1. 证明人 ( )	6. 求职信 ( )
2. 教育背景 ( )	7. 英语水平 ( )
3. 所获荣誉 ( )	8. 个人资料 ( )
4. 求职意向 ( )	9. 成绩单 ( )
5. 实习经历 ( )	10. 工作经历 ( )

Ⅱ. Multiple choice

1. It could be extremely ________ and frustrating when you learned you didn't get the job.
   A. disappointing B. satisfying C. exciting
2. Before going for an interview, it is important to ________ as much information as possible about the company.
   A. take out B. bring away C. look up
3. If you get the job, ________ yourself to it and you will be rewarded.
   A. devoting B. devote C. to devote
4. During a job interview, it's advisable that you talk about your strengths, but you shouldn't avoid mentioning your ________.
   A. hobbies B. accomplishments C. weaknesses
5. We finally managed to make the customers ________ of the quality of the car.
   A. to convince B. convincing C. convinced
6. A college is going to ________ a student's ability based on grades and then make the decision to admit or refuse an applicant.
   A. experiment B. observe C. assess
7. If ________ for the job, you'll be informed soon.
   A. accept B. accepting C. accepted
8. ________ details of this position, please contact our manager Mr. Wang at 5789432.
   A. For B. On C. with

Ⅲ. Fill in the blanks with the proper choices given.

It's difficult to find your dream job. This month, the search will become even harder as new graduates start looking for work. With many companies still struggling, many job hunters have a very difficult task ahead of them.

Don't be disappointed. ___(1)___. Consider using the following three methods to find and get your dream job.

The Front Door

___(2)___. These include checking company websites and employment services. Since lots of people use the front door, make yourself stand out. This means understanding what a company really needs and creatively showing you can meet that need. ___(3)___. Find out who is doing the hiring, and try to give them your resume in person. Then follow up with a phone call, letter or visit.

The Back Door

A back-door job search means finding a job through people you know. The back door can help you avoid the crowd of job applicants using the front door. ___(4)___. What if you don't have any connections at your dream company? Apply for an internship.

The Invisible Door

If you're still not getting hired, you might be running into an invisible（无形的）door. Invisible doors are things companies want from new employees but don't list as job requirements. Many companies have special company cultures. Others are looking for people with very specific skills. ___(5)___.

It's not always easy to find a good job. But with a little work, a door will open!

A. This type of job search uses traditional ways of getting hired

B. If you aren't showing you are a perfect fit, you won't get hired

C. There are still jobs out there for those who know what to look for

D. But you still need to work hard to show you are the best person for the job

E. Read about each company you apply to

GLOSSARY

A

absorb	[əbˈzɔ:b]	*v.*	吸收	Unit 10
academic	[ˌækəˈdemɪk]	*adj.*	学术的	Unit 1
academics	[ˌækəˈdemɪks]	*n.*	学术知识	Unit 11
accomplish	[əˈkʌmplɪʃ]	*v.*	完成	Unit 5
accounting	[əˈkaʊntɪŋ]	*n.*	会计	Unit 1
acesulfame	[əsɪʌlˈfeɪm]	*n.*	安赛蜜	Unit 3
achieve	[əˈtʃi:v]	*v.*	完成	Unit 1
acid	[ˈæsɪd]	*n.*	酸	Unit 4
acidophilus	[ˌæsɪˈdɒfələs]	*n.*	嗜酸菌；乳酸杆菌	Unit 4
acquire	[əˈkwaɪə]	*v.*	获得	Unit 8
actually	[ˈæktʃuəli]	*adv.*	实际上	Unit 1
adequate	[ˈædɪkwət]	*adj.*	足够的	Unit 6
adhere	[ədˈhɪə(r)]	*v.*	遵守	Unit 11
allergen	[ˈælədʒən]	*n.*	过敏原	Unit 3
allergy	[ˈælədʒi]	*n.*	过敏反应	Unit 3
alter	[ˈɔ:ltə]	*v.*	改变	Unit 3
amenity	[əˈmi:nəti]	*n.*	便利设施	Unit 6
anachronism	[əˈnækrənɪzəm]	*n.*	过时事物	Unit 6
announce	[əˈnaʊns]	*v.*	宣布	Unit 9
annoying	[əˈnɔɪɪŋ]	*adj.*	使烦恼的	Unit 1
antibiotic	[ˌæntibaɪˈɒtɪk]	*n.*	抗生素	Unit 3
antioxidant	[ˌæntiˈɒksɪdənt]	*n.*	抗氧化物质	Unit 8
applicant	[ˈæplɪkənt]	*n.*	申请人	Unit 11
apply	[əˈplaɪ]	*v.*	申请	Unit 11
appreciation	[əˌpri:ʃiˈeɪʃn]	*n.*	欣赏	Unit 5
approach	[əˈprəʊtʃ]	*v.*	接近	Unit 2
approve	[əˈpru:v]	*v.*	同意	Unit 10
architectural	[ˌɑ:kɪˈtektʃərəl]	*adj.*	建筑方面的	Unit 10
area	[ˈeəriə]	*n.*	范畴	Unit 2
arrangement	[əˈreɪndʒmənt]	*n.*	布置	Unit 10
artificial	[ˌɑ:tɪˈfɪʃl]	*adj.*	人工的	Unit 3
aspartame	[əˈspɑ:teɪm]	*n.*	阿斯巴甜代糖	Unit 3
aspect	[ˈæspekt]	*n.*	方面	Unit 2

assess	[əˈses]	*v.*	评估	Unit 7
assignment	[əˈsaɪnmənt]	*n.*	任务；作业	Unit 1
associated	[əˈsəʊʃieɪtɪd]	*adj.*	有关联的	Unit 3
assume	[əˈsjuːm]	*v.*	假设	Unit 4
attain	[əˈteɪn]	*v.*	达到	Unit 8
attempt	[əˈtempt]	*v.*	努力	Unit 7
attraction	[əˈtrækʃn]	*n.*	有吸引力的事	Unit 8
auction	[ˈɔːkʃn]	*n.*	拍卖	Unit 2
automated	[ˈɔːtəmeɪtɪd]	*adj.*	自动化的	Unit 4
automatically	[ˌɔːtəˈmætɪkli]	*adv.*	自动地	Unit 4
autonomous	[ɔːˈtɒnəməs]	*adj.*	自主的	Unit 7
B				
bacteria	[bækˈtɪəriə]	*n.*	细菌	Unit 3
barely	[ˈbeəli]	*adv.*	仅仅；刚刚	Unit 1
battery	[ˈbætri]	*n.*	电池	Unit 9
benefit	[ˈbenɪfɪt]	*n.*	优势	Unit 8
berry	[ˈberi]	*n.*	浆果	Unit 10
bid	[bɪd]	*v.*	出价	Unit 10
billion	[ˈbɪljən]	*n.*	十亿	Unit 2
bonus	[ˈbəʊnəs]	*n.*	红利	Unit 6
bookkeeping	[ˈbʊkkiːpɪŋ]	*n.*	记账	Unit 6
boom	[buːm]	*v.*	迅速发展	Unit 2
bouquet	[buˈkeɪ]	*n.*	花束	Unit 10
bowel	[ˈbaʊəl]	*n.*	肠	Unit 4
breed	[briːd]	*n.*	类型	Unit 7
broiler	[ˈbrɔɪlə(r)]	*n.*	肉鸡	Unit 9
bud	[bʌd]	*n.*	花苞	Unit 10
buttermilk	[ˈbʌtəmɪlk]	*n.*	脱脂乳	Unit 4
by-product	[ˈbaɪˌprɒdʌkt]	*n.*	副产品	Unit 8
C				
calorie	[ˈkæləri]	*n.*	卡路里（热量单位）	Unit 3
campus	[ˈkæmpəs]	*n.*	（大学、学院的）校园	Unit 1
canned	[kænd]	*adj.*	罐装的	Unit 3
carbon	[ˈkɑːbən]	*n.*	碳	Unit 5
carnation	[kɑːˈneɪʃn]	*n.*	康乃馨	Unit 10
casual	[ˈkæʒuəl]	*adj.*	非正式的	Unit 11

catapult	[ˈkætəpʌlt]	*n.*	弹射器	Unit 7
category	[ˈkætəgəri]	*n.*	类别	Unit 6
cell	[sel]	*n.*	细胞	Unit 3
challenging	[ˈtʃælɪndʒɪŋ]	*adj.*	挑战性的	Unit 1
chemical	[ˈkemɪkl]	*n.*	化学品	Unit 9
cholesterol	[kəˈlestərɒl]	*n.*	胆固醇	Unit 4
chronic	[ˈkrɒnɪk]	*adj.*	长期的	Unit 7
circulation	[ˌsɜːkjəˈleɪʃn]	*n.*	流通	Unit 2
circumstance	[ˈsɜːkəmstəns]	*n.*	境况	Unit 6
claim	[kleɪm]	*v.*	声称	Unit 3
classically	[ˈklæsɪkli]	*adv.*	最常见地	Unit 4
client	[ˈklaɪənt]	*n.*	客户	Unit 10
climatological	[ˌklaɪmətəˈlɒdʒɪkl]	*adj.*	与气候学有关的	Unit 8
coast	[kəʊst]	*n.*	海岸；海滨	Unit 5
cognitive	[ˈkɒgnətɪv]	*adj.*	认知的	Unit 3
combat	[ˈkɒmbæt]	*v.*	战斗	Unit 9
combine	[kəmˈbaɪn]	*v.*	（使）结合，组合	Unit 4
commerce	[ˈkɒmɜːs]	*n.*	贸易；商务	Unit 2
commercial	[kəˈmɜːʃl]	*adj.*	商业的	Unit 2
commodity	[kəˈmɒdəti]	*n.*	商品	Unit 4
commonplace	[ˈkɒmənpleɪs]	*adj.*	普遍的	Unit 3
communication	[kəˌmjuːnɪˈkeɪʃn]	*n.*	交际	Unit 1
compartment	[kəmˈpɑːtmənt]	*n.*	分隔间；隔层	Unit 4
competition	[ˌkɒmpəˈtɪʃn]	*n.*	比赛；竞争	Unit 1
complementary	[ˌkɒmplɪˈmentri]	*adj.*	相称的；互补的	Unit 10
complex	[ˈkɒmpleks]	*adj.*	复杂的	Unit 5
confidence	[ˈkɒnfɪdəns]	*n.*	自信	Unit 1
confined	[kənˈfaɪnd]	*adj.*	狭窄而围起来的	Unit 4
confirm	[kənˈfɜːm]	*v.*	确认	Unit 6
confusion	[kənˈfjuːʒn]	*n.*	混淆	Unit 2
conscious	[ˈkɒnʃəs]	*adj.*	有意识的	Unit 5
conservation	[ˌkɒnsəˈveɪʃn]	*n.*	保护	Unit 5
considerate	[kənˈsɪdərət]	*adj.*	考虑周到的	Unit 1
consumption	[kənˈsʌmpʃn]	*n.*	（能量、食物或材料的）消耗	Unit 9
contractor	[kənˈtræktə(r)]	*n.*	承包商	Unit 10

conventionally	[kən'venʃənəli]	*adv.*	传统地	Unit 8
convince	[kən'vɪns]	*v.*	使信服	Unit 11
coordination	[kəʊˌɔ:dɪ'neɪʃn]	*n.*	协调	Unit 2
cover	['kʌvə(r)]	*v.*	包括	Unit 2
criterion	[kraɪ'tɪəriən]	*n.*	标准	Unit 5
crucial	['kru:ʃl]	*adj.*	关键性的	Unit 9
cultured	['kʌltʃəd]	*adj.*	培养的	Unit 4
current	['kʌrənt]	*adj.*	现在的	Unit 11
customize	['kʌstəmaɪz]	*v.*	定制	Unit 8
D				
damage	['dæmɪdʒ]	*v.*	损坏	Unit 3
deadline	['dedlaɪn]	*n.*	截止日期	Unit 11
debate	[dɪ'beɪt]	*v.*	辩论	Unit 1
decently	['di:sntlɪ]	*adv.*	体面地	Unit 11
decoration	[ˌdekə'reɪʃn]	*n.*	装饰品	Unit 10
define	[dɪ'faɪn]	*v.*	设定	Unit 2
definition	[ˌdefɪ'nɪʃn]	*n.*	定义	Unit 2
delicacy	['delɪkəsi]	*n.*	审慎；周到	Unit 7
destination	[ˌdestɪ'neɪʃn]	*n.*	目的地	Unit 6
detail	['di:teɪl]	*n.*	细节	Unit 6
deterioration	[dɪˌtɪərɪə'reɪʃən]	*n.*	退化	Unit 8
determined	[dɪ'tɜ:mɪnd]	*adj.*	坚定的	Unit 1
diagnosis	[ˌdaɪəg'nəʊsɪs]	*n.*	诊断	Unit 9
diagnostician	[ˌdaɪəgnɒs'tɪʃən]	*n.*	诊断专家	Unit 9
diarrhea	[ˌdaɪə'ri:ə]	*n.*	腹泻	Unit 4
differentiate	[ˌdɪfə'renʃieɪt]	*v.*	区分	Unit 11
digest	[daɪ'dʒest]	*v.*	消化	Unit 4
digestible	[daɪ'dʒestəbl]	*adj.*	易消化的；口感好的	Unit 4
digestive	[daɪ'dʒestɪv]	*adj.*	消化的	Unit 3
digital	['dɪdʒɪtl]	*adj.*	数码的	Unit 7
digitized	['dɪdʒɪtaɪzd]	*adj.*	数字化的	Unit 2
disability	[ˌdɪsə'bɪləti]	*n.*	障碍	Unit 3
disabled	[dɪs'eɪbl]	*adj.*	残疾的	Unit 6
disadvantage	[ˌdɪsəd'vɑ:ntɪdʒ]	*n.*	缺点	Unit 9
disadvantaged	[ˌdɪsəd'vɑ:ntɪdʒd]	*adj.*	弱势的	Unit 2
disrupt	[dɪs'rʌpt]	*v.*	打乱	Unit 9

distinct	[dɪˈstɪŋkt]	*adj.*	截然不同的	Unit 2
distinguish	[dɪˈstɪŋgwɪʃ]	*v.*	区分	Unit 2
distribution	[ˌdɪstrɪˈbjuːʃn]	*n.*	经销；（网络）分销	Unit 2
diversity	[daɪˈvɜːsəti]	*n.*	多样性	Unit 8
donate	[dəʊˈneɪt]	*v.*	捐赠	Unit 5
driven	[ˈdrɪvn]	*adj.*	奋发努力的	Unit 1
drone	[drəʊn]	*n.*	无人机	Unit 7
drought	[draʊt]	*n.*	久旱	Unit 10
dynamically	[daɪˈnæmɪkəli]	*adv.*	动态地	Unit 10
E				
ecological	[ˌiːkəˈlɒdʒɪkl]	*adj.*	生态的	Unit 5
ecosystem	[ˈiːkəʊsɪstəm]	*n.*	生态系统	Unit 5
efficient	[ɪˈfɪʃnt]	*adj.*	效率高的	Unit 2
electronic	[ɪˌlekˈtrɒnɪk]	*adj.*	电子的	Unit 2
element	[ˈelɪmənt]	*n.*	基本部分	Unit 10
eliminate	[ɪˈlɪmɪneɪt]	*v.*	清除	Unit 9
emission	[iˈmɪʃn]	*n.*	排放	Unit 5
emphasis	[ˈemfəsɪs]	*n.*	强调	Unit 4
emphasize	[ˈemfəsaɪz]	*v.*	强调	Unit 11
engineering	[ˌendʒɪˈnɪərɪŋ]	*n.*	工程	Unit 3
enhance	[ɪnˈhɑːns]	*v.*	提高；增强	Unit 4
enhancer	[ɪnˈhɑːnsə(r)]	*n.*	增强剂	Unit 3
entire	[ɪnˈtaɪə(r)]	*adj.*	整个的	Unit 2
essence	[ˈesns]	*n.*	本质	Unit 1
establish	[ɪˈstæblɪʃ]	*v.*	建立	Unit 5
ethical	[ˈeθɪkl]	*adj.*	道德的	Unit 5
evaluation	[ɪˌvæljuˈeɪʃn]	*n.*	评价	Unit 9
exotic	[ɪgˈzɒtɪk]	*adj.*	异国风情的	Unit 6
exponential	[ˌekspəˈnenʃl]	*adj.*	越来越快的	Unit 9
extant	[ekˈstænt]	*adj.*	现存的	Unit 5
extensive	[ɪkˈstensɪv]	*adj.*	大量的	Unit 10
extracurricular	[ˌekstrəkəˈrɪkjʊlə(r)]	*adj.*	课外的	Unit 11
F				
facility	[fəˈsɪləti]	*n.*	设施	Unit 5
familiarize	[fəˈmɪliəraɪz]	*v.*	使熟悉	Unit 11
feeble	[ˈfiːbl]	*adj.*	虚弱的	Unit 9

ferment	[fə'ment]	*v.*	（使）发酵	Unit 4
fermentation	[ˌfɜːmen'teɪʃn]	*n.*	发酵（作用）	Unit 4
fertilizer	['fɜːtəlaɪzə(r)]	*n.*	肥料	Unit 7
financially	[faɪ'nænʃəli]	*adv.*	经济上；财政上	Unit 2
flavor	['fleivə]	*n.*	味道	Unit 3
fleet	[fliːt]	*n.*	车队；船队	Unit 2
florist	['flɒrɪst]	*n.*	花艺师	Unit 10
fluent	['fluːənt]	*adj.*	流利的	Unit 11
focus	['fəʊkəs]	*v.*	集中（注意力、精力）	Unit 1
folk	[fəuk]	*n.*	人们	Unit 3
form	[fɔːm]	*v.*	形成；构形；组织	Unit 2
foul	[faʊl]	*v.*	弄脏；污染	Unit 10
fragrance	['freɪgrəns]	*n.*	香味	Unit 10
frustrate	[frʌ'streɪt]	*v.*	使懊恼	Unit 4
function	['fʌŋkʃn]	*n.*	功能	Unit 9
fungal	['fʌŋgl]	*adj.*	真菌的	Unit 7
G				
garbage	['gɑːbɪdʒ]	*n.*	垃圾；废物	Unit 5
gene	[dʒiːn]	*n.*	基因	Unit 3
generate	['dʒenəreɪt]	*v.*	产生	Unit 4
genetic	[dʒə'netɪk]	*adj.*	基因的	Unit 3
grain	[greɪn]	*n.*	谷物	Unit 3
grammatically	[grə'mætɪklɪ]	*adv.*	文法上地；语法上地	Unit 11
greenery	['griːnəri]	*n.*	青枝绿叶	Unit 10
groom	[gruːm]	*v.*	打扮	Unit 11
guarantee	[ˌgærən'tiː]	*v.*	保证	Unit 11
guideline	['gaɪdlaɪn]	*n.*	指导方针	Unit 2
H				
harness	['hɑːnɪs]	*v.*	利用	Unit 2
heavily	['hevɪli]	*adv.*	在很大程度上	Unit 3
hen	[hen]	*n.*	母鸡	Unit 9
herb	[hɜːb]	*n.*	草本植物	Unit 8
herd	[hɜːd]	*n.*	牧群	Unit 9
highlight	['haɪlaɪt]	*v.*	强调	Unit 11
hitter	['hɪtə(r)]	*n.*	要员；强有力的人	Unit 11
hormone	['hɔːməʊn]	*n.*	激素	Unit 8

humane	[hju:ˈmeɪn]	*adj.*	人道的	**Unit 4**
humble	[ˈhʌmbl]	*adj.*	不起眼的	**Unit 2**
I				
identify	[aɪˈdentɪfaɪ]	*v.*	发现	**Unit 7**
implement	[ˈɪmplɪmənt]	*n.*	工具	**Unit 2**
impression	[ɪmˈpreʃn]	*n.*	印象	**Unit 11**
incentive	[ɪnˈsentɪv]	*n.*	激励	**Unit 8**
indefinitely	[ɪnˈdefɪnətli]	*adv.*	无限期地	**Unit 8**
infection	[ɪnˈfekʃn]	*n.*	传染	**Unit 9**
inferior	[ɪnˈfɪəriə(r)]	*adj.*	比不上的	**Unit 7**
infestation	[ˌɪnfɛsˈteɪʃən]	*n.*	横行；侵扰	**Unit 7**
inflammatory	[ɪnˈflæmətri]	*adj.*	发炎的；炎性的	**Unit 4**
infrastructure	[ˈɪnfrəstrʌktʃər]	*n.*	基础设施	**Unit 2**
ingredient	[ɪnˈgri:diənt]	*n.*	成分；配料	**Unit 3**
initial	[ɪˈnɪʃl]	*adj.*	最初的	**Unit 10**
initiative	[ɪˈnɪʃətɪv]	*n.*	倡议	**Unit 9**
insect	[ˈɪnsekt]	*n.*	昆虫	**Unit 3**
installation	[ˌɪnstəˈleɪʃn]	*n.*	安装	**Unit 10**
institution	[ˌɪnstɪˈtju:ʃn]	*n.*	机构	**Unit 4**
intellectual	[ˌɪntəˈlektʃuəl]	*adj.*	有才智的	**Unit 1**
interchangeably	[ˌɪntəˈtʃeɪndʒəbli]	*adv.*	可互换地	**Unit 2**
interior	[ɪnˈtɪəriə(r)]	*adj.*	内部的	**Unit 10**
interview	[ˈɪntəvju:]	*n.*	面试	**Unit 11**
intolerance	[ɪnˈtɒlərəns]	*n.*	不容忍	**Unit 4**
intoxicated	[ɪnˈtɒksɪkeɪtɪd]	*adj.*	中毒的	**Unit 9**
involve	[ɪnˈvɒlv]	*v.*	包含	**Unit 2**
irrigation	[ˌɪrɪˈgeɪʃn]	*n.*	灌溉	**Unit 7**
L				
label	[ˈleɪbl]	*v.*	用标签标明	**Unit 8**
lactic	[ˈlæktɪk]	*adj.*	乳的	**Unit 4**
lactose	[ˈlæktəʊs]	*n.*	乳糖	**Unit 4**
landfill	[ˈlændfɪl]	*n.*	废物填埋地	**Unit 5**
landscape	[ˈlændskeɪp]	*n.*	景观	**Unit 10**
lawn	[lɔ:n]	*n.*	草坪	**Unit 7**
league	[li:g]	*n.*	联赛；社团	**Unit 1**
legume	[ˈlegju:m]	*n.*	豆科作物	**Unit 3**

lengthy	[ˈleŋθi]	*adj.*	冗长的	Unit 11
lettuce	[ˈletɪs]	*n.*	莴苣	Unit 7
lily	[ˈlɪli]	*n.*	百合花	Unit 10
link	[lɪŋk]	*n.*	联系	Unit 2
livestock	[ˈlaɪvstɒk]	*n.*	家畜	Unit 8
logistics	[ləˈdʒɪstɪks]	*n.*	物流	Unit 2
lower	[ˈləʊə(r)]	*v.*	减少	Unit 9
luxury	[ˈlʌkʃəri]	*n.*	奢侈的享受；奢侈品	Unit 5
M				
machinery	[məˈʃiːnəri]	*n.*	（统称）机器	Unit 7
maintenance	[ˈmeɪntənəns]	*n.*	维护	Unit 8
make-up	[meɪkʌp]	*n.*	组成成分	Unit 3
manipulation	[məˌnɪpjʊˈleɪʃ(ə)n]	*n.*	操作	Unit 7
manual	[ˈmænjuəl]	*n.*	使用手册	Unit 7
manufacturer	[ˌmænjuˈfæktʃərə(r)]	*n.*	生产商	Unit 5
manure	[məˈnjʊə(r)]	*n.*	粪肥	Unit 4
medication	[ˌmedɪˈkeɪʃn]	*n.*	药物	Unit 9
merely	[ˈmɪəli]	*adv.*	仅仅	Unit 2
millet	[ˈmɪlɪt]	*n.*	小米	Unit 2
mineral	[ˈmɪnərəl]	*n.*	矿物质	Unit 3
minimal	[ˈmɪnɪml]	*adj.*	极少的；最低的	Unit 3
misconception	[ˌmɪskənˈsepʃn]	*n.*	错误认识	Unit 8
mitigate	[ˈmɪtɪgeɪt]	*v.*	减轻	Unit 7
model	[ˈmɒdl]	*n.*	模式	Unit 2
modify	[ˈmɒdɪfaɪ]	*v.*	修改	Unit 3
mold	[məʊld]	*n.*	霉菌	Unit 4
monitor	[ˈmɒnɪtə(r)]	*v.*	监控	Unit 7
moral	[ˈmɒrəl]	*adj.*	道德上的	Unit 1
moreover	[mɔːrˈəʊvə(r)]	*adv.*	此外	Unit 9
motivate	[ˈməʊtɪveɪt]	*v.*	激励	Unit 1
multiple	[ˈmʌltɪpl]	*adj.*	多种多样的	Unit 8
multispectral	[ˌmʌltiˈspektrəl]	*adj.*	多光谱	Unit 7
mutation	[mjuːˈteɪʃn]	*n.*	（生物物种的）变异	Unit 9
N				
navigation	[ˌnævɪˈgeɪʃn]	*n.*	导航	Unit 7
negative	[ˈnegətɪv]	*adj.*	有害的	Unit 3

neotame	[niə'teɪm]	*n.*	纽甜	Unit 3
nervous	['nɜ:vəs]	*adj.*	神经系统的	Unit 9
nutrient	['nju:triənt]	*n.*	营养素	Unit 8
nutritional	[nju'trɪʃənl]	*adj.*	营养的	Unit 3
nutritious	[nju'trɪʃəs]	*adj.*	有营养的	Unit 2
O				
objective	[əb'dʒektɪv]	*adj.*	客观的	Unit 9
obliged	[ə'blaɪdʒd]	*adj.*	感激的	Unit 11
optimal	['ɒptɪməl]	*adj.*	最佳的	Unit 8
optimize	['ɒptɪmaɪz]	*v.*	使最优化	Unit 8
organic	[ɔ:'gænɪk]	*adj.*	有机的	Unit 8
organism	['ɔ:gənɪzəm]	*n.*	有机体；生物	Unit 5
otherwise	['ʌðəwaɪz]	*adv.*	另外	Unit 4
overcome	[ˌəʊvə'kʌm]	*v.*	克服	Unit 1
overlap	[ˌəʊvə'læp]	*v.*	部分重叠	Unit 2
overwhelming	[ˌəʊvə'welmɪŋ]	*adj.*	无法抗拒的	Unit 1
P				
packaging	['pækɪdʒɪŋ]	*n.*	包装	Unit 2
parameter	[pə'ræmɪtə]	*n.*	范围；规范	Unit 2
participate	[pɑ:'tɪsɪpeɪt]	*v.*	参加	Unit 5
pasture	['pæstʃə(r)]	*n.*	牧场	Unit 7
perception	[pə'sepʃn]	*n.*	认知	Unit 7
perform	[pə'fɔ:m]	*v.*	履行	Unit 6
periodically	[ˌpɪərɪ'ɒdɪkəli]	*adv.*	定期地	Unit 7
persistent	[pə'sɪstənt]	*adj.*	持续的	Unit 3
personal	['pɜ:sənl]	*adj.*	个人的	Unit 1
personality	[ˌpɜ:sə'næləti]	*n.*	性格	Unit 11
personalize	['pɜ:sənəlaɪz]	*v.*	使个性化	Unit 11
pesticide	['pestɪsaɪd]	*n.*	杀虫剂	Unit 8
phase	[feɪz]	*n.*	阶段	Unit 10
pose	[pəuz]	*v.*	造成	Unit 3
poultry	['pəʊltri]	*n.*	家禽	Unit 9
precise	[prɪ'saɪs]	*adj.*	准确的	Unit 2
precision	[prɪ'sɪʒn]	*n.*	准确	Unit 8
preliminary	[prɪ'lɪmɪnəri]	*adj.*	初步的	Unit 10
premium	['pri:miəm]	*n.*	额外费用	Unit 4

presentation	[ˌprezn'teɪʃn]	*n.*	提交；报告	Unit 11
preservation	[ˌprezə'veɪʃn]	*n.*	保存	Unit 2
preservative	[prɪ'zɜːvətɪv]	*n.*	防腐剂	Unit 3
primarily	[praɪ'merəli]	*adv.*	主要地	Unit 8
priority	[praɪ'ɒrəti]	*n.*	首要事情	Unit 1
probation	[prə'beɪʃn]	*n.*	试读期	Unit 1
proceed	[prə'siːd]	*v.*	接着做	Unit 10
process	['prəʊses]	*n.*	过程	Unit 2
process	['prəʊses]	*v.*	加工	Unit 2
procurement	[prə'kjʊəmənt]	*n.*	采购	Unit 2
produce	['prɔdjuːs]	*n.*	农产品；产品	Unit 2
product	['prɒdʌkt]	*n.*	产品	Unit 3
productive	[prə'dʌktɪv]	*adj.*	多产的	Unit 7
profession	[prə'feʃn]	*n.*	职业	Unit 7
professor	[prə'fesə(r)]	*n.*	教授	Unit 1
prolong	[prə'lɒŋ]	*v.*	延长	Unit 3
prominent	['prɒmɪnənt]	*adj.*	重要的	Unit 10
protest	[prə'test]	*v.*	抗议	Unit 5
pulp	[pʌlp]	*n.*	纸浆	Unit 5
punctuality	[ˌpʌŋktjʊ'ælɪti]	*n.*	准时	Unit 11
purchase	['pɜːtʃəs]	*v.*	买	Unit 2
Q				
qualify	['kwɒlɪfaɪ]	*v.*	符合；使合格	Unit 5
quantity	['kwɒntəti]	*n.*	数量	Unit 8
R				
rag	[ræg]	*n.*	破布	Unit 5
range	[reɪndʒ]	*n.*	（变动或浮动的）范围	Unit 9
rapidly	['ræpɪdli]	*adv.*	快速地	Unit 2
rash	[ræʃ]	*n.*	皮疹	Unit 3
ratio	['reɪʃiəʊ]	*n.*	比例	Unit 8
realistic	[ˌriːə'lɪstɪk]	*adj.*	现实的	Unit 1
recipe	['resəpi]	*n.*	秘诀	Unit 11
recreation	[ˌrekri'eɪʃn]	*n.*	娱乐	Unit 6
recycle	[ˌriː'saɪkl]	*v.*	回收利用	Unit 5
reference	['refrəns]	*n.*	证明信；证明人	Unit 11
release	[rɪ'liːs]	*v.*	发布	Unit 2

remarkable	[rɪˈmɑːkəbl]	*adj.*	非凡的	Unit 2
request	[rɪˈkwest]	*v.*	要求	Unit 6
reservation	[rezəˈveɪʃ(ə)n]	*n.*	预定	Unit 6
residential	[ˌrezɪˈdenʃl]	*adj.*	住宅的	Unit 10
resilient	[rɪˈzɪliənt]	*adj.*	有适应力的	Unit 3
resistance	[rɪˈzɪstəns]	*n.*	抵抗力	Unit 3
resolve	[rɪˈzɒlv]	*v.*	解决	Unit 2
restore	[rɪˈstɔː(r)]	*v.*	恢复	Unit 9
retailer	[ˈriːteɪlə]	*n.*	零售店	Unit 2
reusable	[ˌriːˈjuːzəbl]	*adj.*	可重复使用的	Unit 5
ripped	[rɪpt]	*adj.*	撕破的	Unit 5
rot	[rɒt]	*v.*	腐烂	Unit 8
rotary	[ˈrəʊtəri]	*adj.*	旋转的	Unit 7
rural	[ˈrʊərəl]	*adj.*	农村的	Unit 2
S				
saccharin	[ˈsækərɪn]	*n.*	糖精	Unit 3
salary	[ˈsæləri]	*n.*	薪金	Unit 11
satisfied	[ˈsætɪsfaɪd]	*adj.*	满意的	Unit 11
secondary	[ˈsekəndri]	*adj.*	次要的	Unit 9
sector	[ˈsektə]	*n.*	行业	Unit 2
security	[sɪˈkjʊərəti]	*n.*	安全	Unit 6
seed	[siːd]	*n.*	籽	Unit 3
semester	[sɪˈmestə(r)]	*n.*	学期	Unit 1
sense	[sens]	*n.*	道理	Unit 1
sensor	[ˈsensə(r)]	*n.*	传感器	Unit 7
sheer	[ʃɪə(r)]	*adj.*	完全的	Unit 4
shortage	[ˈʃɔːtɪdʒ]	*n.*	不足	Unit 7
sign	[saɪn]	*v.*	签字	Unit 6
single	[ˈsɪŋg(ə)l]	*adj.*	单一的	Unit 6
slaughter	[ˈslɔːtə]	*v.*	屠宰	Unit 8
sloppiness	[ˈslɒpɪnəs]	*n.*	马虎	Unit 11
sourcing	[ˈsɔːsɪŋ]	*n.*	寻源采购；得到供货	Unit 2
specialized	[ˈspeʃəlaɪzd]	*adj.*	专门的	Unit 2
specific	[spəˈsɪfɪk]	*adj.*	具体的	Unit 1
specifically	[spəˈsɪfɪkli]	*adv.*	具体地；特别地	Unit 11
spectrum	[ˈspektrəm]	*n.*	光谱	Unit 7

spread	[spred]	*v.*	传播	Unit 3
stained	[steɪnd]	*adj.*	沾有污渍的	Unit 5
standard	[ˈstændəd]	*n.*	标准	Unit 5
stem	[stem]	*n.*	（花草的）茎	Unit 10
storage	[ˈstɔːrɪdʒ]	*n.*	贮存	Unit 2
strain	[streɪn]	*n.*	（植物、动物的）品种	Unit 9
strength	[streŋθ]	*n.*	长处	Unit 1
strengthen	[ˈstreŋθn]	*v.*	加强	Unit 2
stretch	[stretʃ]	*n.*	一片	Unit 5
subcategory	[sʌbˈkætɪgəri]	*n.*	子范畴	Unit 2
subsidy	[ˈsʌbsədi]	*n.*	补贴	Unit 4
substrate	[ˈsʌbstreɪt]	*n.*	基底	Unit 8
sucralose	[ˈsuːkrələʊs]	*n.*	三氯蔗糖	Unit 3
superbug	[ˈsuːpəbʌg]	*n.*	超级细菌	Unit 9
supervision	[ˌsuːpəˈvɪʒn]	*n.*	监督	Unit 10
survive	[səˈvaɪv]	*v.*	挺过；存活	Unit 1
sustainability	[səsˌteɪnəˈbɪlɪti]	*n.*	可持续性	Unit 8
sustainable	[səˈsteɪnəbl]	*adj.*	可持续的	Unit 4
switch	[swɪtʃ]	*v.*	转变	Unit 4
symptom	[ˈsɪmptəm]	*n.*	症状	Unit 4
synthetic	[sɪnˈθetɪk]	*adj.*	合成的	Unit 8
systemic	[sɪˈstiːmɪk]	*adj.*	系统的	Unit 9
T				
take-off	[ˈteɪkɒf]	*n.*	（飞机）起飞	Unit 7
target	[ˈtɑːgɪt]	*v.*	把……作为攻击目标	Unit 9
technology	[tekˈnɒlədʒi]	*n.*	技术	Unit 2
term	[tɜːm]	*n.*	词语	Unit 3
terminology	[ˌtɜːmɪˈnɒlədʒi]	*n.*	术语	Unit 6
texture	[ˈtekstʃə(r)]	*n.*	质地；口感	Unit 4
theme	[θiːm]	*n.*	主题	Unit 10
threat	[θret]	*n.*	威胁	Unit 5
top-notch	[ˌtɒpˈnɒtʃ]	*adj.*	卓越的	Unit 11
total	[ˈtəʊtl]	*v.*	总数达	Unit 2
touch	[tʌtʃ]	*n.*	细微之处	Unit 10
toxin	[ˈtɔksɪn]	*n.*	毒素	Unit 3
track	[træk]	*v.*	跟踪	Unit 9

traditional	[trəˈdɪʃənl]	*adj.*	传统的	Unit 4
transaction	[trænˈzækʃn]	*n.*	（一笔）交易	Unit 2
transfer	[trænsˈfɜː(r)]	*v.*	转移	Unit 9
transit	[ˈtrænzɪt]	*n.*	运输	Unit 2
trash	[træʃ]	*n./v.*	垃圾；弄脏；弄乱	Unit 5
trend	[trend]	*n.*	趋势	Unit 6
trillion	[ˈtrɪljən]	*n.*	万亿	Unit 2
typist	[ˈtaɪpɪst]	*n.*	打字员	Unit 11
U				
ultimate	[ˈʌltɪmət]	*adj.*	最终的	Unit 2
umbrella	[ʌmˈbrelə]	*n.*	综合体	Unit 2
undeniable	[ˌʌndiˈnaiəbl]	*adj.*	不可否认的	Unit 3
unicorn	[ˈjuːnɪkɔːn]	*n.*	（传说中的）独角兽	Unit 11
unique	[juˈniːk]	*adj.*	独特的	Unit 6
uptake	[ˈʌpteɪk]	*n.*	吸收	Unit 8
urogenital	[ˌjʊərə(ʊ)ˈdʒenɪt(ə)l]	*adj.*	泌尿生殖器的	Unit 9
V				
value	[ˈvæljuː]	*n.*	价值	Unit 3
variation	[ˌveəriˈeɪʃn]	*n.*	变化	Unit 7
vastly	[ˈvɑːstli]	*adv.*	非常	Unit 4
veggie	[ˈvedʒi]	*n.*	蔬菜	Unit 3
vehicle	[ˈviːəkl]	*n.*	交通工具	Unit 5
ventilation	[ˌventɪˈleɪʃən]	*n.*	通风	Unit 8
venue	[ˈvenjuː]	*n.*	聚会地点	Unit 11
vertical	[ˈvɜːtɪkl]	*adj.*	垂直的	Unit 8
virus	[ˈvaɪrəs]	*n.*	病毒	Unit 3
visual	[ˈvɪʒuəl]	*adj.*	视觉的	Unit 7
vitamin	[ˈvɪtəmɪn]	*n.*	维生素	Unit 3
W				
weakness	[ˈwiːknəs]	*n.*	弱点	Unit 1
width	[wɪdθ]	*n.*	宽度	Unit 7
Y				
yeast	[jiːst]	*n.*	酵母；酵母菌	Unit 4
yield	[jiːld]	*n.*	产量	Unit 7

Phrases & Expressions

A

a strain of	一种	Unit 9
abnormal amino acid	异常氨基酸	Unit 3
access to	使用……的机会（权利）	Unit 9
according to	根据；按照	Unit 6
adhere to	遵循	Unit 11
agricultural drone	农业无人机	Unit 7
antibiotic resistance	抗生素抗药性	Unit 9
apart from	除了……外(还)	Unit 1
apply for	申请	Unit 11
apply...to	应用于……	Unit 8
artificial intelligence（AI）	人工智能	Unit 9
artificial sweetener	人工甜味剂	Unit 3
as a result	结果	Unit 2
as well	也	Unit 2
ASAP	尽快	Unit 11
at an angle	斜的	Unit 10
at least	至少	Unit 4
attempt to do	尝试做某事	Unit 7
automatic sprayer	自动喷洒机	Unit 7

B

balance out	（使）平衡；（使）相抵	unit 10
band together	联合	Unit 5
battery chicken	电池鸡（层架式鸡笼所养的鸡）	Unit 9
be assigned to	被指定；被分配	Unit 6
be confined to	局限于……	Unit 9
be considered as	被认为；看作是	Unit 6
be enthusiastic about	对……热心	Unit 11
be harmful to	对……有害	Unit 5
be included in	包括在	Unit 6
be inferior to	逊色于	Unit 7
be involved in	涉及；参与	Unit 6
be linked to	与……有关	Unit 3
be referred to	被提及；涉及	Unit 6
be related to	与……有关	Unit 2

become involved in	参与	Unit 2
bed former	作床机	Unit 7
benefit from	受益于	Unit 4
big data	大数据	Unit 2
Brexit	英国脱欧	Unit 7
brighten up	为……增辉添彩	Unit 10
broiler chicken	肉鸡	Unit 9
C		
cater to	迎合	Unit 5
check in	入住	Unit 6
clear-cut	明显的	Unit 9
cloud computing	云计算	Unit 2
coco peat	椰纤土	Unit 8
cold chain logistics	冷链物流	Unit 2
combine harvester	联合收割机	Unit 7
combined... with...	与……结合	Unit 4
compared with	与……相比	Unit 9
contribute to	有助于……	Unit 1
cut down	削减	Unit 5
D		
data set	数据集	Unit 9
deal with	涉及；处理	Unit 2
deli meat	熟食肉	Unit 3
depend on (upon)	取决于；依赖	Unit 4
digestive system	消化系统	Unit 9
disguise as	伪装	Unit 3
dispose of	处理	Unit 5
drought tolerant and water-saving garden	抗旱节水型花园	Unit 10
due to	由于	Unit 8
E		
encourage sb. to do	鼓励某人做某事	Unit 5
end up	以……告终	Unit 1
even though	即使	Unit 3
F		
fall under	被归入	Unit 2
familiarize...with...	使某人熟悉某物	Unit 11
figure out	想出	Unit 11
fill out	填写	Unit 11

focus on	集中于	Unit 2
for the sake of	出于……考虑	Unit 8
free-range chicken	散养鸡	Unit 9
G		
genetic mutation	基因突变	Unit 9
Global Positioning System	全球定位系统	Unit 7
go hand in hand	相辅相成；密切相关	Unit 6
H		
have a reservation	有预定	Unit 6
have access to	有机会或权利使用	Unit 8
have problems with...	在……方面有问题	Unit 1
hay rake	搂草机	Unit 7
herb garden	草本园	Unit 10
I		
immigration policy	移民政策	Unit 7
in addition to	除……之外	Unit 4
in charge of	掌管	Unit 11
in connection with	与……有关	Unit 8
in contrast with	与……相比	Unit 4
in essence	实质上	Unit 1
in general	总的来说	Unit 5
in most cases	大多数情况下	Unit 11
in order to	为了	Unit 3
in the eyes of	在……看来	Unit 5
in the long term	从长远来看	Unit 8
in the wake of	随着……而来	Unit 7
inbound and outbound freight	进出港货运	Unit 2
interact with	与……互动	Unit 5
J		
jump to conclusions	妄下结论	Unit 9
L		
lactic acid bacteria	乳酸菌	Unit 4
large-scale	大规模的	Unit 9
laying hen	蛋鸡	Unit 9
lead to	导致	Unit 3
lean meat	瘦肉	Unit 3
lie in	在于……	Unit 10
light cream	淡奶油	Unit 4

line ripper	垄行成型机	Unit 7
live streaming	直播	Unit 2
look for	寻找；寻求	Unit 6
look out for	留意	Unit 11
M		
make sense	有道理	Unit 1
migrant workers	外来务工者	Unit 7
monosodium glutamate	味精	Unit 3
more or less	或多或少	Unit 6
more than	不仅仅是	Unit 8
move away from	远离	Unit 8
N		
natural-style garden	自然风格花园	Unit 10
nervous system	神经系统	Unit 9
neurological system	神经系统	Unit 3
no longer	不再	Unit 5
nutritional content	营养含量	Unit 3
O		
omega-3 fatty acid	ω-3不饱和脂肪酸	Unit 8
on one's left	在某人左手边	Unit 6
on the second floor	在二楼	Unit 6
P		
participate in	参加	Unit 1
pay attention to	重视；对……注意	Unit 10
people with disabilities	残疾人	Unit 6
pile up	（使）成堆	Unit 5
pipe... in	用管道传送……	Unit 9
place emphasis on	重视	Unit 5
play a role in	在……中起作用	Unit 10
prevent... from...	阻止	Unit 11
put simply	简单地说	Unit 2
R		
refer to	指的是	Unit 2
refined carbohydrate	精制碳水化合物	Unit 3
regardless of	不论	Unit 8
result from	由……产生；由……引起	Unit 4
result in	导致	Unit 5
retail sales	零售额	Unit 2

reverse shipping	逆向运输	Unit 2
rotary hoe	旋转锄地机	Unit 7
S		
scale...up	扩大	Unit 8
set aside	把……放置一旁	Unit 11
set...up	为……做准备	Unit 1
shelf life	保质期	Unit 3
sign up	报名（参加课程）	Unit 1
so-called	所谓的	Unit 9
sodium nitrite	亚硝酸钠	Unit 3
sour cream	酸奶油（烹饪用）	Unit 4
stand for	是……的缩写；代表	Unit 2
stem from	根源是	Unit 2
supply chain management	供应链管理	Unit 2
support facility	辅助设施	Unit 2
T		
take off	（飞机）起飞	Unit 7
tend to	往往会；常常	Unit 3
thanks to	由于	Unit 2
the Internet of Things	物联网	Unit 2
the number of	……数量	Unit 6
the pros and cons	事物的利与弊	Unit 3
to a certain degree	在某种程度上	Unit 2
trans fat	反式脂肪	Unit 3
translate ...into...	把（思想、感情等）用另一种形式表达出来	Unit 10
transportation industry	交通运输业	Unit 2
try one's best	尽力	Unit 11
turn to...for help	向……寻求帮助	Unit 1
U		
unmanned aerial vehicle	无人机	Unit 7
urogenital system	泌尿生殖系统	Unit 9
V		
vertical farm	垂直农场	Unit 7
W		
wake up	醒来	Unit 11
water-themed garden	水系主题花园	Unit 10
when it comes to sth.	当涉及某事时	Unit 3

REFERENCES

[1] https://wenku.baidu.com/view/0698ab83ccbff121dd3683e0.html

[2] http://talk.kekenet.com/show_2432

[3] https://www.thoughtco.com

[4] https://www.renrendoc.com/p-49521942.html

[5] https://www.zujuan.com/question/detail/7565361

[6] https://marketbusinessnews.com/financial-glossary/e-commerce/amp/

[7] https://www.parklu.com

[8] https://www.wisegeek.com

[9] https://www.pcmag.com/reviews/norton-antivirus-plus

[10] https://us.norton.com/internetsecurity-malware-what-is-antivirus.html

[11] http://tinyurl.com/yacv2qql

[12] https://www.who.int

[13] https://www.conceptlifesciences.com/capability/pesticide-residue-testing/

[14] https://www.voxnature.com/top-7-most-dangerous-foods-and-why-you-should-avoid-them/

[15] https://www.symptomfind.com/nutrition-supplements/pros-and-cons-of-genetically-modified-foods/

[16] https://en.angelyeast.com/products/yeast-baking-ingredients/a-800-bread-improver.html

[17] https://m.alibaba.com/product/62532094005/Bread-Baking-Machine-Stainless-Steel-32.html

[18] https://www.heavenlyhomemakers.com/summer-activities-with-the-kids-go-berry-apple-or-peach-picking

[19] https://traveltips.usatoday.com/types-rooms-hotels-2980.html

[20] https://www.city-of-hotels.com/165/hotel-services-business.html

[21] http://5utk.ks5u.com/main.aspx?mod=ext&ac=paper&op=search#sou_suo_hou_a

[22] https://www.k-line.net.au/products/speedtiller/

[23] https://www.hisour.com/agricultural-drone-40802/?share=telegram&nb=1

[24] https://www.roboticsbusinessreview.com/unmanned/agriculture-robots-four-global-trends-watch/

[25] https://www.farmliner.com.au/

[26] https://www.jobmonkey.com/farming/farm-equipment/

[27] https://www.mdpi.com/2304-8158/2/4/488/htm

[28] http://www.israelagri.com/?CategoryID=435&ArticleID=665

[29] https://www.redtone.com/smartfarming/smart-greenhouse/

[30] https://www.mdpi.com/2304-8158/2/4/488/htm

[31] http://smithmeadows.com/farm/what-is-free-range-chicken/

[32] http://news.cri.cn/gb/5324/2004/12/30/982%40407438.htmhttps://www.peta.org/features/case-controlled-atmosphere-killing/
[33] http://language.chinadaily.com.cn/2016-09/19/content_26816811.htm
[34] http://www.hxen.com/englisharticle/yingyuyuedu/2019-08-20/522883.html
[35] https://draxe.com/nutrition/health-benefits-of-eggs/
[36] https://ahdb.org.uk/knowledge-library/african-swine-fever
[37] https://www.wikihow.com
[38] http://talk.oralpractice.com/list_2_33_0_1.html
[39] http://www.orwiglandscapearchitect.com/home_tour_multilevel.htm
[40] https://www.chinahighlights.com/travelguide/architecture/features-garden.htm
[41] https://www.fromhousetohome.com/garden/pick-the-right-plants
[42] https://careertu.com/blog/whatsyourbiggestweakness/
[43] https://paper.i21st.cn/m/story/55687.html
[44] https://jamybechler.com/7-tips-for-applying-for-your-next-job/?utm_sq=fck313ldmc&utm_source=Twitter&utm_medium=social&utm_campaign=ULeadership101&utm_content=Blog+Posts
[45] https://www.whatcareerisrightforme.com/blog/5-helpful-tips-to-make-a-good-first-impression-at-a-job-interview/
[46] https://www.artofmanliness.com/articles/how-to-dress-for-job-interview/
[47] http://gzyy.cooco.net.cn/testdetail/331920/
[48] 朱兴全. 兽医专业英语［M］. 北京：中国农业出版社，2003.